ETHNOGRAPHIC CONSTRUCTIONS OF INDIGENOUS OTHERS

This book examines the ways in which indigeneity interacts with climate change politics at multiple levels and at the same time offers a self-critical reflection on the role of ethnographic research (and researchers) in this process. Through a multi-sited ethnography, it shows how indigeneity and climate change mitigation are at this point so intensely intertwined that one cannot be clearly understood without considering the other. While indigenous identities have been (re)defined in relation to climate change, it argues that Indigenous Peoples continue to subvert pervasive notions of the nature/culture dichotomy and disrupt our understanding of what it means to be human in relation to nature. It encourages students and researchers in anthropology, international development, and other related fields to engage in more meaningful reflection on the epistemic shortcomings of "the West", including in our own research, and to acknowledge the ongoing role of power, coloniality, extractivism, and whiteness in climate change discourses.

George Byrne is a researcher at the London School of Economics and Political Science in the Centre for Women, Peace and Security. His PhD in International Development at the University of Sussex focussed on power and identities in international development, particularly the position of Indigenous Peoples and the role of indigeneity in climate change negotiations. George also holds an MSc in Social Research Methods, an MA in International Relations, and a BA in Latin American Development Studies and Spanish. His current research focusses on reflexive methodologies and the role of power in the interpretation of empirical research.

ETHNOGRAPHIC CONSTRUCTIONS OF INDIGENOUS OTHERS

Indigeneity, Climate Change, and the Limits of Western Epistemology

George Byrne

Routledge
Taylor & Francis Group
NEW YORK AND LONDON

Designed cover image: © Getty Images |oxygen

First published 2024
by Routledge
605 Third Avenue, New York, NY 10158

and by Routledge
4 Park Square, Milton Park, Abingdon, Oxon, OX14 4RN

Routledge is an imprint of the Taylor & Francis Group, an informa business

ISBN: 978-1-032-37776-6 (hbk)
ISBN: 978-1-032-37777-3 (pbk)
ISBN: 978-1-003-34186-4 (ebk)

DOI: 10.4324/9781003341864

Typeset in Sabon LT Pro
by KnowledgeWorks Global Ltd.

CONTENTS

LIST OF ABBREVIATIONS

AbC	Accumulation by Conservation
AIDESEP	Interethnic Association for the Development of the Peruvian Forest *Asociación Interétnica de Desarrollo de la Selva Peruana*
APA	Association of Amerindian Peoples of Guyana
CBA	Cost Benefit Analysis
CBD	Convention on Biological Diversity
CfRN	Coalition for Rainforest Nations
CIDOB	The Confederation of Indigenous Peoples of the Bolivian Orient *Confederación de Pueblos Indígenas de Bolivia*
COICA	Coordinator of Indigenous Organizations of the Amazon River Basin *Coordinadora de las Organizaciones Indígenas de la Cuenca Amazónica*
CONIVE	Indian National Council of Venezuela *Consejo Nacional Indio de Venezuela*
CONFENIAE	Confederation of Indigenous Nationalities of the Ecuadorian Amazon *Confederación de Nacionalidades Indígenas de la Amazonía Ecuatoriana*
CONAIE	Confederation of Indigenous Nationalities of Ecuador *Confederación de Nacionalidades Indígenas del Ecuador*
CO_2	Carbon Dioxide

COP	Conference of the Parties to the UNFCCC
EDF	Environmental Defense Fund
FCPF	Forest Carbon Partnership Facility
FOAG	Federation of Indigenous Organizations of Guyana *Fédération des Organisations Autochtones de Guyane*
FOIN	Federation of Indian Organizations of the Napo *Federacion de Organizaciones Indígenas del Napo*
FONAKISE	Federation of Kichwa Organisations of Sucumbíos-Ecuador *Federación de Organizaciones de la Nacionalidad Kichwa de Sucumbìos Ecuador*
FPIC	Free, Prior and Informed Consent
GEG	Global Environmental Governance
GOE	Government of Ecuador *Gobierno del Ecuador*
IEN	Indigenous Environmental Network
IGO	Inter-governmental Organisation
IIPFCC	International Indigenous Peoples' Forum on Climate Change
IISD	International Institute for Sustainable Development
ILO	International Labour Organization
INDC	Intended Nationally Determined Contribution
IPCC	Intergovernmental Panel on Climate Change
IPCCA	Indigenous Peoples' Biocultural Climate Change Assessment Initiative
IPO	Indigenous Peoples' Organisation
MAE	Ecuadorian Ministry of the Environment *Ministerio del Ambiente d Ecuador*
MALI	Lima Museum of Art *Museo de Arte de Lima*
MBI	Market-based Instrument
MINAM	Peruvian Ministry of the Environment *Ministerio del Ambiente de Peru*
NGO	Non-governmental Organisation
NORAD	Norwegian Agency for Development Cooperation
NPA	Natural Protected Area
OIS	Organization of Indian Surinam
ONIC	Organization of Indigenous Nationalities of Colombia *Organización Nacional Indígena de Colombia*
OPIAC	National Organisation of Indigenous Peoples of the Colombian Amazon

	Organización de los Pueblos Indígenas de la Amazonía Colombiana
ORPIA	Venezuelan Regional Organisation of Indigenous Peoples of the Amazon *Organización Regional de los Pueblos Indígenas de Amazonas*
PES	Payment for Ecosystem Services
RAISG	Amazon Network of Georeferenced Socio-Environmental Information *Red Amazónica de Información Socioambiental Georreferenciada*
RAN	Rainforest Action Network
RCP	Rio Conventions Pavillion
REDD(+)	Reducing Emissions from Deforestation and Forest Degradation (plus)
RIA	Amazon Indigenous REDD+ *REDD+ Indígena Amazónico*
SDSN	Sustainable Development Solutions Network
SSE	Sustainable Stock Exchanges Initiative
tCO_2e	Tonnes of Carbon Dioxide Equivalent
TNC	The Nature Conservancy
UK	United Kingdom
UN	United Nations
UN-REDD	United Nations Programme on Reducing Emissions from Deforestation and Forest Degradation
UNCCD	United Nations Convention to Combat Desertification
UNCED	United Nations Conference on Environment and Development
UNCSD	United Nations Conference on Sustainable Development
UNCTAD	United Nations Conference on Trade and Development
UNDP	United Nations Development Programme
UNDPHDR	United Nations Development Programme Human Development Report
UNDRIP	United Nations Declaration on the Rights of Indigenous Peoples
UNEP	United Nations Environment Programme
UNESCO	United Nations Educational, Scientific and Cultural Organization
UNFCCC	United Nations Framework Convention on Climate Change
UNI	Union of Indigenous Nations of Brazil *União das Nações Indígenas*

UNORCID	United Nations Office for REDD+ Coordination in Indonesia
UNPFIP	United Nations Permanent Forum for Indigenous Peoples
WHRC	Woods Hole Research Center
WWF	World Wide Fund for Nature

ACKNOWLEDGEMENTS

This book is filled with the energy and influence of more people than I could possibly acknowledge, and there are also some who cannot be mentioned for reasons of anonymity. So, I wish to begin by expressing my gratitude to all those, named and unnamed, who have contributed to my research. In particular, I would like to thank the many Indigenous people who granted me access to their workspaces, communities, homes, and lives, as well as those who I encountered only briefly. These interactions formed the foundation of my research, and it would have been impossible without them. The staff of COICA were exceptionally generous in this regard, allowing me to hang around for months on end, asking the same questions over and over again, and letting me tag along with them to various conferences and meetings. They also included me in their social events and celebrations, which made me feel less like I was intruding and left me with some wonderful memories.

From within the academic space, I would first like to thank the supervisors of my doctoral studies, Dr Evan Killick and Professor James Fairhead. They trusted me to follow some unusual paths and allowed me to experiment with a variety of methods, and they pushed me gently when I needed it. I hope I didn't cause them too much stress and frustration over the years. Two more people who had a significant influence on my work are Dr Juli Hazlewood, who steered me in the direction of decolonial methodologies while I was in Ecuador, and Dr Lindsay Ofrias, who shared her knowledge of similar places and experiences. This led to conversations about our research, as well as our experiences of trauma and fear, that were filled with empathy, kindness, and compassion. When I returned from fieldwork, my mind was awash with thoughts and feelings, but in writing I felt paralyzed by anxiety. At a moment when I was utterly lost, Professor Gurminder Bhambra held a series of seminars on decolonial and postcolonial perspectives, which introduced me to a body of literature and a way of thinking and writing that helped me find a voice.

On a personal level, I feel lucky to have had the seemingly endless and unwavering support and encouragement from a lot of people. My mum, dad, brothers, and sister haven't always understood what I have been doing for all these years, but they have supported me and have been proud of me regardless. When I needed people who did understand the experience of feeling at odds with the academic world, I turned to other doctoral students, particularly Dr Yusef Bakkali and Dr Yeyang Su, and to people who had been through it all before, including Professor Tamsin Bradley and Dr Zara Ramsay. They made me feel as if I was not alone and reminded me that this type of work is hard for everyone. When I needed to escape my work, my friends were always there. In Ecuador I had Peter, Amanda, Kirsty, and Ben, and when I came home, Trevor, Rachel, Gareth, B, Ellen, Dan, Tuci, and many more. My friend Angela also opened her home to me, providing me with the quiet space and time that I needed, allowing me to immerse myself in the writing process.

All of these people are present in my writing, but more than anyone else my wife, Jo, has made this book possible. She has made significant personal and professional sacrifices over the years, and yet, her patience, resilience, love, kindness, and compassion seem to know no bounds. I am so grateful to her for this, and in many ways, this book is as much a product of her effort as it is my own. I simply could not have done it without her.

The research for this book was joint funded by the UKRI Economic and Social Research Council (ESRC) 1+3 PhD studentship programme and The University of Sussex Doctoral Training Centre.

INTRODUCTION

In 2014, feeling frustrated by the evident failures of governments and international organisations, particularly the United Nations (UN), to effectively address global environmental concerns, I set out to better understand how Indigenous Peoples might contribute to the fight against climate change. For a while, I had been aware of the belief held by some academics and activists that indigenous knowledges might offer potential alternatives to dominant understandings of our relationship with the environment, and, while I was somewhat sceptical of this, I remained hopeful that someone somewhere was thinking and acting differently. A part of me was, and still is, drawn to the idea that if we in "the West" can only reconnect with nature and reimagine our relationships with the earth, we can also become stewards of the planet: maybe, if we take climate change and other environmental risks seriously, we can move towards a future characterised by hope and care rather than fear and domination. In this book, I reflect on this theme with reference to my interactions with Indigenous Peoples in spaces ranging from a small indigenous settlement in the Ecuadorian Amazon to the indigenous political and activist spaces at UN climate change conferences.

The empirical research for this book was mostly conducted between March 2014 and December 2015, and while the experiences I had during this time made writing this book possible, my fieldwork, which began as part of a PhD project, did not go as planned. It was coloured by failure and frustration, and by seeing, hearing of, and experiencing violence and fear. My research caused harm – certainly to me, and most likely to others – and

DOI: 10.4324/9781003341864-1

though this is probably a common outcome, it is rarely discussed openly in academic literature. Such negative consequences are, however, far from irrelevant in shaping the landscape of research: for me, they shaped where and how I would conduct my fieldwork, which in turn led me to reflect on how the knowledge I was extracting and (re)producing was both subject to and implicated in a process of selection. The result of this is that the Indigenous people who are most often "given voice", both in academic research and in climate change negotiations, necessarily exhibit particular characteristics: they can be identified as indigenous, they are accessible, and they must, to some degree, be cooperative with non-indigenous academic researchers and development practitioners.

With this problem in mind, this book explores the ways in which conducting research can contribute to the creation of categories of people – Others – whose presence in our work plays a complicated role. On the one hand, it can facilitate indigenous representation and participation, while on the other it can serve to legitimise and strengthen existing patterns of social, cultural, and epistemological dominance. I reflect on how the research process itself prefigures and reconfigures relationships and how it changes the people who encounter it or engage with it. This includes the researcher and consenting participants, but it also has implications for those who are excluded or who choose not to take part. My observations are presented through an ethnographic, as well as a partly autoethnographic, account of various interactions between a variety of actors around the broad themes of indigeneity and climate change, and it is intended as an illustration of the thinking, writing, and reflection processes. For this reason, methodological considerations, some of which might otherwise have been omitted, remain in this book because without this it wouldn't necessarily be obvious to the reader why I made certain decisions relating to methods, style, and content.

To begin with, for example, it is worth saying that when I began the research for this book, I hadn't intended to write extensively about methodology at all. Rather, I had planned a fairly typical anthropological research project, in which I would spend a year conducting a study of an Amazonian indigenous settlement in Ecuador. In doing so, I expected to discover how they understood and interacted with the climate change mitigation strategies that affected them and their territory. But I didn't conduct the traditional long-term ethnographic field study I had proposed, and I was instead led by various circumstances to unexpected fieldsites where I found an indigeneity that was different to what I had imagined. As my own needs changed, particularly regarding personal security and time commitments, I also changed the parameters of where I was willing to work, and which people would fulfil the role of Other in my work. The people I found in these alternative places were easily identified as indigenous (by me), they were accessible (to me), and they were cooperative (with me), but above all, the places and spaces in which I chose to work were safer (for me). This particular group of Indigenous people and this particular way of *being Indigenous* were shaped by and for the places in which I encountered them, and for me they were just-Other-enough. This optimal degree of Otherness meant they were sufficiently different to be interesting, but not so different that they became too difficult to work with.

Even before this, my selected fieldsites had been influenced by fairly arbitrary circumstances. I had chosen to conduct my research in Ecuador because, from 2009 to 2010, I had worked for a year in Quito as an intern for an international environmental organisation called Amazon Watch. Our small team shared an office and much of our day-to-day work with the lawyers representing the plaintiffs in a major environmental lawsuit, which had originally been filed in 1993 against Chevron (then Texaco). On visits to the Amazon region, I saw some of the contamination that had been caused by oil extraction and the devastating impact this had on the local people. During this time, I also became familiar with the many barriers to environmental justice in Ecuador and with the national and regional indigenous environmental movement, which was entangled with struggles for Indigenous Peoples' rights, land, and identities. These conflicts were also being shaped by the broader political and social changes that were taking place in Latin America at the time, particularly the Pink Tide; a shift towards socialism, which had characterised the politics of the region in the past decade or so and had been accompanied by a growing anti-imperialist and decolonial sentiment.

When I returned to Ecuador almost five years later as a PhD student, I did so with good intentions. I wanted to use the opportunity I now had to elevate indigenous voices and maybe, in some small way, make a positive contribution to their ongoing struggle. I wasn't explicitly aware of it at the time, or at least had not realised what it meant, but I later came to understand that I had also intended to take something from the people I encountered: their knowledge and presence in my work and my life was (and still is) *valuable* to me. Once I began trying to work directly with Indigenous Peoples, I quickly realised that I had underestimated the role of relational encounters, power, and identities (not least my own as a white, male academic from the United Kingdom) in research, particularly when someone like me is working with marginalised or "hard to reach" communities. I became very uncomfortable with the way I had designed my project: from the very beginning, I had selected questions, methods, and a methodology that were interesting to me and were deemed appropriate within British academic culture. It had to be this way, of course, because if it had not been, I would not have received the funding, support, or ethical clearance to begin my research in the first place. My carefully considered and well-meaning plans, however, did not align with the realities of the field.

As my perspective shifted, I became less concerned with specific elements of the climate change mitigation strategies that were being implemented in indigenous territories in Ecuador – such as REDD+ and Socio Bosque (see Chapter 1) – and the impact they were having on Indigenous Peoples at the local level. While they remain and are still present in this book, these strategies became a lens through which to explore notions of indigeneity in relation to climate change more broadly and to consider the role of power in reifying – or "making up" to use Hacking's (1986) term – categories of people and designating them as legitimised Others. Equally, my plan to spend extensive time with a community of the Sápara Indigenous people, who are located in the east of Ecuador's Amazon region and across the border with Peru, also didn't work out: my interactions with them were complicated and challenging from the start, and I don't think I could have continued my research with them without causing considerable harm. Nonetheless, the time I did spend with them was a critical moment that abruptly shifted my understanding of my identity in relation to my research as well as changing my perspective on research in general. Thus, the ethnographic chapters that focus on this short time mostly reflect on what I now consider to be a period of problematic and failed research. Yet it was also the source of valuable lessons, for me at least, and part of what this book aims to do is help other researchers avoid making the same mistakes.

The shifts between various research sites and themes allow for this book to be linked to the theoretical approaches of anthropology of globalisation, development anthropology, and global ethnography (see, e.g., Burawoy, 2000; Edelman & Haugerud, 2005; Tsing, 2005). That said, in taking a reflexive position and applying a critical perspective to my own work, I intend to contribute to a process of "anthropologising the West", as Rabinow put it (1986, cited in Banerjee, Chio, & Mir, 2009), though I do so very much from an insider perspective. My hope is that this will shed light on some of the social and economic structures, and the institutional and institutionalised practices, that give rise to the inequitable, and unjust, power relations that are pervasive in research processes and in development outcomes (Banerjee et al., 2009, p. 3). Nonetheless, I am aware that this requires the reader to begin with a certain capacity to cope with the inevitable cognitive dissonance that emerges out of this approach (Alvesson & Sköldberg, 2017, p. 314). There is a messiness to writing about these conflicting social relations from the subjective perspective of a researcher, and while the friction this creates can energise the text, it can also generate tensions and inconsistencies (see Tsing, 2005).

Such tensions can become magnified in the writing process when the authors thoughts and feelings are misaligned. As Adrienne Rich reflected in the 1984 essay *Notes Towards a Politics of Location*:

> I do not any longer believe – my feelings do not any longer allow me to believe – that the white eye sees from the centre. Yet I often find myself thinking as if I still believed that were true. Or rather, my thinking stands still. I feel in a state of arrest, as if my brain and heart were refusing to speak to each other (Rich, 1984, p. 226).

A central thread that runs through this book is that it endeavours to work through and push beyond this state of arrest. I have sought to illustrate and recognise inconsistencies in my own thinking and to be open about the complexity of doing this type of work. Because of this, what I offer in this book could perhaps be described as a post-modern ethnography *of* development anthropology; it is a partial narrative consisting "of fragments of discourse intended to evoke in the minds of both reader and writer an emergent fantasy of a possible world of common-sense reality, and thus provoke an aesthetic integration that will have a therapeutic effect" (Tyler, 1986, p. 125). I recognise, however, that any such positive outcome is limited in its scope, and the main beneficiary of writing about my experiences in this way is, of course, me. So, while on the one hand this book is still positioned at the point where ideas around indigeneity and climate change meet in academic research, at its core it is a methodological and ethical reflection on *why, and for whom, this type of research is conducted.*

With these questions in mind, I reflect upon and try to make explicit some of the ways in which knowledge, particularly regarding identities, is produced and how this is linked to power dynamics in the worlds of development and research. This can be understood as a general process of selection whereby particular knowledges and ways of being are negotiated, reified, and legitimised. In relation to being *Indigenous*, this happens in a multitude of interconnected places, including spaces where climate change is being discussed and in the processes and products of academic research, as well as across the boundaries between them. I try to make the strands of this web of power unambiguous in my writing, which includes not shying away from my own identity and its role in shaping my encounters and the research I conducted. Because of this, I have become a somewhat central character of this book, which is as uncomfortable and off-putting for me as it likely is for the reader. Nonetheless, I couldn't find a way to communicate what I have learnt through my research without writing openly about myself. The reasons for this are two-fold. The first is because, through taking indigenous, decolonial, and feminist scholarship *seriously* (which I haven't always done), I gradually became aware of how impossibly intertwined my research is with my experiences, my personality, and my emotions. The second reason is because I can't say much about what it means to be indigenous, other than how it appears to me, which will only make sense to the reader if you know how I arrived at these views. So, in order to engage meaningfully with these ideas, I write myself in where it feels appropriate, discussing how my identity, my personal experiences, and my emotions shape my work. My hope is that this will encourage other scholars to engage in similar practices and to understand them as an ethical necessity, regardless of whether such reflections appear in their final research outputs or not. Ideally, this will happen long before making the mistakes I did and will raise a crucial question in the mind of any researcher when considering similar work: *should I do it at all?*

I don't want to oversimplify the relationship between researchers and the communities we write about, or to suggest that all research of this kind necessarily has negative outcomes. As such, it is important to state that Western knowledge and ways of being – including ways of being Others – are not simply forced upon one group of individuals by another: it is more complex than that. Many varied actors in encounters that take place from the interpersonal level to the international political level shape which questions are asked and by whom as well as the ways in which they can and cannot be answered. In relation to this book, this means that the people with whom I worked are implicitly legitimised as being indigenous by my choice to write about them, and my work is also legitimised by their presence in it. But, of course, there are many more people being indigenous in myriad ways and in a multitude of places, most of which I have never seen. I therefore offer only a partial and quite problematic perspective, which itself serves as an example of the very process that I am attempting to illustrate: being indigenous does not simply refer to being part of a category of people who happen to be indigenous *to* somewhere any more than being non-indigenous refers to someone *not* being from anywhere. Rather, being indigenous (or not) emerges out of an ongoing struggle to "make up", and make use of, particular notions of indigeneity. This struggle, within which a complex web of actors seeks to assert their agency, is shaped by power, which is diffuse and flows through society in all directions and to all extremities (see Foucault, cited in Fraser, 1981, p. 272). Yet the historically constituted asymmetries that exist (Levin, Pratto, & Sidanius, 2006, p. 272) – in power, in access to resources, and in the positive and negative outcomes of development – mean that these negotiations around indigenous and non-indigenous identities are never entirely equitable.

Through this process, contemporary notions of indigeneity have emerged alongside Western fantasies of "sustainable development" (see

Tsing, 2007): both have been reified over time and have become increasingly intertwined. Dominant approaches to climate change mitigation – including various market-based instruments (MBIs) such as REDD+ and other payment for ecosystem services and "offsetting" mechanisms (see Chapter 1) – and indigenous alternatives to them are all positioned within this particular political context and epistemological framework. They are underpinned by the assumptions: 1. That economic development is good and necessary, 2. That development *can* be sustainable, 3. That sustainable development can be achieved through governance at the global level, and 4. That Indigenous Peoples and their knowledges are different but useful in the process. Similarly, my research and this book is positioned within its own epistemological limitations, bounded by a history of Eurocentrism that has resulted in certain ways of being and knowing, i.e., those that emanate from Western formalised knowledge institutions, being situated at the apex of a hierarchy of epistemological communities (Tatman, 2001, p. 138) while the knowledges and cultures of Indigenous Peoples, along with other marginalised and minoritised groups, have been and continue to be subjugated (Akena, 2012, p. 600). Thus, it is necessary to acknowledge that, while this book is critical of it, my research was conducted from and for an institution that is embedded within, as well as being implicated in the preservation of, the institutional logics of a remnant structure of colonialism: British academia.

It is interesting to note that when I began the research for this book, decolonial theory was far from mainstream in international development and anthropology. It seems like the world was a very different place in 2014, and in the time between then and now movements such as Black Lives Matter and Me Too, the increasing urgency of the climate crisis and, of course, the Covid-19 pandemic have all contributed to reconfiguring Western social consciousnesses in various ways. While decoloniality has been somewhat mainstreamed during this time, particularly in UK universities (see, e.g., Hall, Ansley, & Connolly, 2023), and while feminism has seen a popular resurgence, there have also been plenty of backlashes against both. What seems odd to me now, though, is that as a scholar of international development I began thinking about indigeneity and climate change long before I started engaging meaningfully with decolonial or feminist methodologies, and part of what this book offers is an illustration of how I learned to think differently about these themes. Of course, I have also changed during this time, and this means I have had to resist heavily rewriting the observations I made then to reflect what I think and feel now, because this would not be honest. As such, some of my wording and the thoughts I share in the ethnographic and autoethnographic sections of this book are not written exactly as I would express them if I were to start again, and I hope they are taken with this in mind.

This gap I found in my thinking also makes me cautious of what other shortcomings I might see in my current perspective a few years from now, but on a more general level the shifting intellectual landscape in which this book has been written shows how climate change mitigation and research about it can act as central sites of epistemic struggle. Since climate change mitigation policies often pertain to indigenous territories, and because indigenous knowledges are afforded a unique status in some parts of climate change discourse, research and negotiations relating to the environment have become unique arenas in which certain assumptions and epistemological hierarchies can, at least in theory, be challenged by Indigenous Peoples.

Chapter Structure

This book is divided into three parts – **PART I**, *Background and Methodology*, **PART II**, *Seeing and Being Indigenous*, and **PART III**, *Fieldwork that Hurts* – each of which has two chapters. The six chapters all consider indigenous identity, and particularly what it means to be indigenous when climate change is being discussed, and they also reflect upon my own experience *of* my research. I begin with a discussion of the relationship between Indigenous Peoples, development, and "the environment", which is followed by a methodological reflection on how, and indeed *if*, research of this kind could be conducted ethically. I then trace this idea through my fieldsites in a broadly reverse chronological order, which leads the reader from the international to the regional/national, the local, and ultimately to the personal level. This structure also allows me to follow a path inward from the theoretical and methodological considerations (Chapters 1 and 2) to empirical evidence based upon interviews and participant observation (Chapters 3–5), and finally to a reflection on experiential forms of knowledge (Chapter 6).

I have chosen to present the material in this way because, in a sense, that is how it happened. What I mean by this is that in the process of writing I found myself with an underlying feeling for what was important, but without the ability to articulate it clearly. In order to make sense of what I wanted to say and to illustrate it, I needed to trace my ideas back through from the present to the times and spaces where they had arisen. Thus, the empirical chapters begin in a place that is culturally, temporally, and geographically relatively near to where I was when I began writing (a conference space in Paris in 2015), and then they travel to spaces that are more distant (conferences and offices in Latin America, and an indigenous community in the Amazon). Finally, I write about a place that is familiar only to me, and which is also the most difficult to articulate: my own affective experience.

PART I – Chapter 1, *The Emergence of REDD+, Indigenous Alternatives, and Indigenous Resistance*, first introduces the UN Framework Convention on Climate Change (UNFCCC) Conference of the Parties (COP) events and describes how the Reduction of Emissions due to Deforestation and Forest Degradation (UN-REDD) programme and REDD+ emerged out of them. I briefly discuss some of the ways in which notions of "indigenous" and "the environment" have come to be aligned in climate change discourse, particularly in relation to "ecological nobility" and "stewardship" of the rainforests. This was my entry point into researching the role of Indigenous Peoples in climate change mitigation because in these spaces the intention to make use of indigenous knowledges for the goal of sustainable development is made explicit. This is followed by an introduction to the two groups of Indigenous people with whom I interacted most during my research: members Coordinator of Indigenous Organizations of the Amazon (COICA) and representatives of the Sápara Nation of Ecuador.

COICA has become a prominent indigenous voice at the global level and has led the development of an indigenous alternative to REDD+. It is among the most professionalised and well recognised indigenous peoples' organisations (IPOs) and has gone to great lengths to work within the discourse of and institutions of global sustainable development. Conversely, the Sápara people, though they often have representatives at UN conferences, have had a complicated history with REDD+, including rejecting Ecuador's readiness programme: Socio Bosque. They are an example of a more resistant indigeneity that tends to see outside incursions on indigenous territories, including in the name of environmentalism, as a form of ongoing colonisation. This background information is provided to shed some light on the historically constituted structural asymmetries that exist on cultural interfaces where the role of indigeneity in climate change mitigation is being negotiated, and to raise the question of how meaningful climate change negotiations that supposedly include Indigenous Peoples can be when such asymmetries exist. It is certainly not the intention of this chapter to provide a definition of the term "indigenous", but instead to consider how the category has come to be understood and positioned in the political landscape, and ultimately to be employed as a discursive tool in debates and negotiations around climate change.

Chapter 2, *Taking Up Space and Making Up Others*, describes and reflects on the methodological approaches I adopted during the research, data analysis, and writing process of this book. As noted earlier, my original research proposal imagined a relatively "traditional" approach to working with Indigenous Peoples. I planned to go to the rainforest, conduct semi-structured interviews and participant observation over an extended period with a remote, culturally distinct people, and come back having learned something of how "they" understand and interact with externally driven projects in their territories, such as those elaborated under the rubric of REDD+. But throughout my fieldwork I became somewhat disillusioned regarding my own research (as well as anthropology and development studies as fields of research), and my existing misgivings regarding the role of academia in general were compounded.

In questioning how, and indeed *if*, "we" can say anything authoritative, useful, or even innocuous, about what it means to be indigenous I began to draw on feminist and decolonial perspectives to critique my research design. Through these critical lenses, I highlight some of the epistemological and institutional limitations that should be taken into consideration when reading this book. This required me to do some difficult work, particularly in the process of reflecting on how my identity and experiences shape my research and writing. I hope that by describing how I worked my way back through my research this chapter acts as an example of how writing can be approached as a personal process of learning and unlearning while being both critical of and compassionate to oneself and others. I also discuss the style in which the subsequent chapters are written, which is at times unconventional, and choices I made on language and terminology. This includes explaining why I use the term "indigenous" at all and why it is sometimes capitalised and sometimes not, and why I occasionally use the terms we" when referring to non-indigenous people (specifically, Western researchers) and "they" when referring to Indigenous Peoples respectively.

PART II, *Seeing and Being Indigenous*, begins with the first of four ethnographic chapters that together offer an account of my clumsy attempts at participant observation and interviews, which were made across multiple fieldsites between March 2014 and December 2015. Chapter 3, *Negotiating Indigeneity at the Indigenous Pavilion*, focusses on the last of these fieldsites (chronologically speaking), which was the Indigenous Pavilion at COP21, though I also discuss some relevant observations from the Pavilion at the conference, COP20, in Lima in 2014. I present a selection of vignettes that illustrate my impression of this indigenous space located in a non-indigenous place. These accounts are intended to highlight the sometimes-contradictory ways in which indigeneity appears. The chapter opens space to begin considering how this small enclave of indigeneity at COP and these few people, who ostensibly represent millions of Indigenous Peoples around the world, come to represent the totality of what it means to be *Indigenous* here. How and why were these ways of being selected, what does this mean for the concept of indigeneity elsewhere, and how does this identity function in these places where the lives and territories of Indigenous Peoples are being negotiated?

I also intend in this chapter to draw the reader into my own feelings of uneasiness and cynicism in these spaces. This discomfort emerges from a tension between two contradictory ideas relating to indigenous representation: on the one hand, it is a crucial part of the indigenous struggle for self-determination, but on the other hand the few indigenous voices that are selected and legitimised at the Pavilion in turn legitimise the space itself. This is further complicated by the ways in which what counts as *Indigenous* are negotiated, presented, and consumed in the space.

PART II – Chapter 4, *Encountering the Optimal Other*, continues back through my fieldsites, shifting from an international non-indigenous place to a space that geographically and culturally bridges the gap between indigenous and non-indigenous worlds. I first encountered COICA during COP20 in Lima in December 2014 at the first Indigenous Pavilion. From there, I began spending time with members of the organisation in their head office in Quito, Ecuador, and in other locations around the city. This chapter depicts the office and the people in it and recounts a number of events where outside actors (indigenous and non-indigenous) came into contact with COICA, including the organisation's 30th anniversary events; a meeting between COICA, Ecuadorian IPOs, and UN and Government representatives; and a media training workshop conducted by a communications organisation for COICA staff. I consider the ways in which indigeneity appears in these spaces and how it is different to what I observed at the COP events.

PART III, *Fieldwork that Hurts*, is characterised by a move towards a more geographically, culturally, and temporally distant place, and also towards a more reflexive methodological approach. In Chapter 5, *Failures and Frustration during Research Encounters*, I recount a struggle that I encountered early in my fieldwork. I had planned to work with the Sápara People, an Indigenous Nationality in Ecuador, which began with began with two short research trips to a small, isolated community in the Eastern region of the country's Amazon rainforest. It was during this time that I began to realise that the ethical and methodological considerations I had made prior to travelling to Ecuador had been entirely insufficient. I present examples from the time spent in and with this indigenous community, including reflections on workshops I conducted on behalf of a Peruvian NGO and other workshops I observed being conducted by an Ecuadorian NGO. I also discuss on a trip where I accompanied a representative of the community to a conference in Lima, which I found frustrating and difficult to navigate, and I describe my experience of a dispute about money that led to the eventual breakdown of my relationship with my gatekeepers to the community. The intention of this chapter is to offer an honest account of my experience of failed research and to reflect on what went wrong and why. While I hope I caused little harm during this time, it was here that I began to question whether I should be doing this type of research at all.

The final empirical chapter of this book, Chapter 6 – *Fear, the Real, and the Other*, draws the narrative into an explicitly subjective context. I use this chapter to position my work alongside the "post-modern challenge" to development (Gardner & Lewis, 1996, pp. 22–23), recounting and exploring personal experiences in the field, the impact these experiences had on me as a researcher and as a person, and how this shaped my writing. I intend to highlight flaws in my own thinking and research practices, but I also push back against recent Western reimaginings of a benevolent sustainable-developmentalism that continue to assume development will be *good* if it is only done *better*. It includes reflecting on trauma and fear, which is not something that I was able to do in the way I would typically write. As such, I take a more creative approach, using a more "free voice" with the intention of evoking (or provoking) an emotional connection with the reader (Tyler, 1986, p. 125), which at times means I write with less regard for the rigid structures of "good" academic scholarship. The chapter seeks to critique the role of research and to make space for "alternate" forms of knowledge (experiential knowledge, embodied knowledge, intuition, and indigenous knowledges) to be taken seriously within discussions about climate change and in academic research.

The conclusion of this book, *Re-Membering Research*, reflects on the process of bringing together a complex set of experiences and observations to communicate what I have learnt in an intelligible way without simply reducing it to a set of answers to questions and policy recommendations. I do reconsider and provide some thoughts on (though not definitive answers to) the questions that guided the research for this book, including what, if anything, might Indigenous Peoples "know" about climate change that "we" (in the so-called developed nations) do not? I also make some tentative suggestions relating to how a person embarking on similar research might overcome, or at least be better prepared for, some of the problems that I encountered. More importantly, however, I draw together the various threads that run through this book relating to identities, power, and knowledge, to encourage the reader to reflect on why we *do* research and how (and indeed if) it can be done without doing harm to ourselves and others. This is particularly relevant when, as is often the case in anthropology and development studies, we are speaking and writing about themes that relate to unequal distributions of power because the disciplines are themselves implicated in (re)creating and sustaining the asymmetries between researchers and the subjects of our work.

References

Akena, F. A. (2012). Critical analysis of the production of Western knowledge and its implications for indigenous knowledge and decolonization. *Journal of Black Studies, 43*(6), 599–619.

Alvesson, M., & Sköldberg, K. (2017). *Reflexive methodology: New vistas for qualitative research.* London: Sage.

Banerjee, S. B., Chio, V. C., & Mir, R. (2009). The imperial formations of globalization. In S. B. Banerjee, V. C. Chio, & R. Mir (Eds.), *Organizations, markets and imperial formations: Towards an anthropology of globalization* (pp. 3–14). Cheltenham: Edgar Elgar.

Burawoy, M. (Ed.). (2000). *Global ethnography: Forces, connections, and imaginations in a postmodern world.* Los Angeles: University of California Press.

Edelman, M., & Haugerud, A. (Eds.). (2005). *The anthropology of globalization and development: From classical political economy to contemporary neoliberalism.* London: Wiley.

Fraser, N. (1981). Foucault on modern power: Empirical insights and normative confusions. *Praxis International, 1*(3), 272–287.

Gardner, K., & Lewis, D. (1996). *Anthropology, development and the post-modern challenge.* London: Pluto Press.

Hacking, I. (1986). Making up people. In D. E. Wellbery, M. Sosna, & T. C. Heller (Eds.), *Reconstructing individualism* (pp. 222–236). Stanford: Stanford University Press.

Hall, R., Ansley, L., & Connolly, P. (2023). Decolonising or anti-racism? Exploring the limits of possibility in higher education. *Teaching in Higher Education*, 1–17.

Levin, S., Pratto, F., & Sidanius, J. (2006). Social dominance theory and the dynamics of intergroup relations: Taking stock and looking forward. *European Review of Social Psychology, 17*(1), 271–320.

Rich, A. (1984). Notes toward a politics of location. In A. Rich (Ed.), (1994) *Blood, bread, and poetry: Selected prose 1979–1985.* New York: Norton.

Tatman, L. (2001). *Knowledge that matters: A feminist theological paradigm and epistemology.* London: Sheffield Academic Press.

Tsing, A. L. (2005). *Friction: An ethnography of global connection.* Oxford: Princeton University Press.

Tsing, A. L. (2007). Becoming a tribal elder, and other fantasies of green development (reprint). In M. Dove & C. Carpenter (Eds.), *Environmental anthropology: A historical reader.* Malden, MA: Blackwell.

Tyler, S. A. (1986). Post-modern ethnography. In J. Clifford & G. Marcus (Eds.), *Writing culture: The poetics and politics of ethnography* (pp. 122–140). London: University of California Press.

PART I

Background and Methodology

1

THE EMERGENCE OF REDD+, INDIGENOUS ALTERNATIVES, AND INDIGENOUS RESISTANCE

Introduction

In 2016, an article entitled "Indigenous Peoples Central to Climate Action" was published on the website of the United Nations Framework Convention on Climate Change (UNFCCC). It stated that forest communities, including 70 million Indigenous Peoples who depend on those forests for their livelihoods, are "increasingly offering their generations of traditional knowledge to further science". It described them as "stewards of the world's most valuable remaining ecosystems", and argued that

> Indigenous communities across the world have a key role to play in tackling climate change and in achieving the climate goals set out in the historic Paris Climate Change Agreement. Indigenous peoples not only have a right to education, they can educate other groups in sustainable life-styles.
>
> *(UNFCCC, 2016)*

This type of statement is today quite mainstream in climate change discourse, but it is a useful starting point for this chapter because it illustrates a shift in sustainable development thinking that has occurred over the past four decades. Indigenous Peoples were not always described in this way; they were once assumed to be passive "beneficiaries" of development, later were reimagined as "stakeholders", and then gradually came to be recognised as important actors in environmental negotiations. Their knowledge has now become valuable to science, and engaging Indigenous Peoples is

DOI: 10.4324/9781003341864-3

understood as a central component of combatting climate change. However, in environmental negotiations this group of people who were once seen (by Western eyes, at least) as barely human at all – as "savages" to be civilised – has now acquired a status far beyond that of beneficiaries or actors: Indigenous Peoples are hailed as the so-called stewards of the planet[1] and the global *leaders* in the fight against climate change.

When I began researching this theme, I did so with the aim of understanding what, if anything, is actually unique about the knowledges held by Indigenous Peoples, how they might contribute to sustainable development, and particularly how this relates to the commodification of nature. This chapter introduces a number of themes that are related to this question and that are referred to throughout the rest of the book. I begin with some observations from one of my fieldsites, the UNFCCC-Conference of the Parties (COP)21, drawing attention to the contradictory ways in which putting a price on nature is imagined by different actors at the event. I then discuss how REDD+, The UN's voluntary climate change mitigation framework that facilitates various green finance mechanisms, emerged out of the UNFCCC process. I also provide an overview of some of the criticisms and concerns related to these types of approaches, particularly those that pertain to Indigenous Peoples and their territories. The final part of this chapter introduces the two groups of Indigenous people with whom I spent time in Ecuador and elsewhere: the Coordinator of Indigenous Organizations of the Amazon Basin (COICA) and the Sápara, an Indigenous nationality whose territory is partly located in Ecuador. Prior to when I met them, these two groups had each interacted with and understood REDD+ in different ways. COICA had taken a leading role in developing what they described as an indigenous "alternative" to REDD+ called REDD+ Indígena Amazónico (RIA), which was designed to work alongside or within the existing UN framework. The Sápara people I met, however, had been more sceptical of and resistant to such approaches, and they had begun to reject proposals to incorporate their territory into the national REDD+ readiness programme called Socio Bosque.

Visions of REDD+ at COP21

In the Climate Generations Space at COP21 in Paris, 2015, two Pavilions were positioned within earshot of one another, only a few metres apart: one was the International Indigenous Peoples Pavilion,[2] and the other was the Rio Conventions Pavilion.[3] In the early afternoon of Monday, December 7, Satya Tripathi, the Director of The United Nations Office for REDD+ Coordination in Indonesia (UNORCID), arrived at the Rio Pavilion for a panel on *Embedding the Economic Rationale in REDD+*

National Strategies.[4] Tripathi was late and was supposed to be speaking first, and as he rushed to his seat, he was handed the microphone to introduce himself. He opened with the following statement:

> Sorry for being late, I was away at the Sustainable Stock Exchanges Initiative meeting that was going on at the Peninsula Hotel, where the leaders of the stock exchanges across the world were discussing how to move forward given what is happening in Paris. Allow me also to bring you some insights from that, as they are very pertinent. The consensus across the room there is that *you cannot manage what you cannot measure* [emphasis added]. And that is what the valuation of ecosystem services is for.

It is no coincidence that the "consensus" Tripathi described emanated from a Sustainable Stock Exchanges Initiative (SSE)[5] luncheon or that it had been held 20 km away in the thousand-dollar-a-night Peninsula Hotel. His comments highlight a key characteristic of contemporary discourses on sustainable development, which is often *treated* as a consensus: that measurement, valuation, and trade are at its core.

The panel was part of the Rio Pavilion thematic day on REDD+, and it was followed by two more that discussed "Where and How REDD+ can Deliver most Benefits" and "Recent Innovations in Private Finance". The International Institute for Sustainable Development (IISD), a think tank that reports on international environmental negotiations, summarised REDD+ day at the Rio Pavilion as follows:

> ... participants heard from a panel discussion addressing how [economic dimensions] can be underlying drivers to deforestation, and how to understand the economic value of forest ecosystems in the national economy. On how REDD+ can deliver maximum benefits, participants considered ways to assess the potential benefits from REDD+, with examples of cost-benefit analyses (CBA) and mapping exercises being highlighted. The day's final session assessed the trends and developments in the private finance sector over recent years, and addressed the different mechanisms for obtaining private finance. The session also looked at the roles blended finance, institutional investors, and private financiers are currently playing in the land use space.
>
> *(IISD, 2015)*

I spent most of the day at these sessions listening to people speaking (mostly in English) about the many measured and imagined benefits of this economic approach to forests. At the reception that followed, I stood among smartly dressed "professional"-looking people, who were drinking wine and eating cheese, and I spoke with the only person I had noticed criticising the underlying assumptions of REDD+ in his questions directed at the panel. Like me, he was a white male academic from the United Kingdom who also worked with an environmental Non-Governmental Organisation (NGO), but he was a bit older than me, and he also had a much better grasp of the economic jargon. "What they really mean", he said, "is what you haven't measured, you can't trade".

The day before, I had attended an action on the Seine, which had been planned months before. A group of Indigenous people paddled kayaks along the river, while protestors who accompanied them held signs and hung banners from the bridge that read "Living Forests: Indigenous Territories Without Oil", "Defend Sacred Forests and Water", and "No REDD: Now They Want to Sell the Air". Since their inception, numerous Indigenous Peoples Organisations (IPOs)[6] and Environmental NGOs[7] have ardently opposed REDD+ as well as other similar Payment for Ecosystem Services (PES) and carbon trading approaches. These supposedly sustainable *solutions* to climate change and environmental concerns are, in certain circumstances, positioned alongside and critiqued in the same way as other detrimental outcomes of development, including deforestation and water contamination, and it is this that initially drew my attention to the question of how and why Indigenous Peoples do or do not become participants in such initiatives. Standing on the bridge above the flotilla, a representative of the Indigenous Environmental Network (IEN), an IPO from the United States that focusses on issues of environmental and economic justice for Indigenous Peoples, offered me some insight into this:

> I can see how it can be tantalising to some leaders when they see these programmes like REDD and carbon markets, where large sums of money are being dropped into your community. We as Indigenous People understand the struggle, but me personally, I feel we have to take a stand. There is a certain amount of morality in this responsibility to Mother Earth, and we can't cross that line. [...] the REDD programme being proposed that is impacting folks in the Congo and displacing communities there is happening so that they continue to do extraction up in Northern Alberta in the tar sands. So, when you start going into arguments that the market will save the world, you automatically, you have to be sceptical.
>
> [...]
>
> This firm belief in the system and our inability to challenge it is what's causing the problem and so its systems change that we have to see, it's not a matter of coming up with an idea of what new form of capitalism

is going to save the world, because really we have to look at ... how do we switch it up, turn it back around and renew the sacred relationship to Mother Earth? How do we renew the spiritual integrity of the water and Mother Earth and our relationship to it? That's the core message that we are trying to bring.

The opposing viewpoints regarding REDD+ expressed at the Rio Pavilion and the action on the Seine illustrate some of the questions and themes that motivated the research that underpins this book. Of particular interest is that there appears, at first sight, to be an unbridgeable gap between competing perspectives on the role of forests and how to approach them, and particularly the way in which this is (or is not) connected to economics. While one group discusses "the economy", "value", "cost benefit analyses", "finance", and "investors", the other raises questions about "integrity", "morality", and "responsibility" to "Mother Earth" and to other people, and it is highly sceptical of markets. But despite this apparent dichotomy, some groups have sought a middle ground.

On REDD+ day at COP, over in the Indigenous Pavilion at precisely the same time as the panels referred to above, REDD+ was being discussed differently. The panel organised by COICA was considering local experiences from across the Amazon and presenting mitigation and adaptation proposals from various Indigenous people, which included REDD+ Indígena Amazónico (RIA), a project that seeks to ease the friction between the "conventional" REDD+ vision and the Indigenous Peoples' movement by incorporating indigenous "cosmovisions" *into* a project that is strategically aligned with, but distinct from, REDD+. The Indigenous Pavilion at COP is the main location of the first ethnographic chapter of this book (see Part II, Chapter 3), but understanding the role of this space at the event first requires some consideration of how REDD+ emerged out of the UNFCCC process, how it has involved Indigenous Peoples, and why it has at times been resisted.

The UNFCCC and COP

The UNFCCC was adopted in New York in May 1992, and the following month, the convention was opened for signatures at the United Nations Conference on Environment and Development (UNCED) in Rio de Janeiro. It was known as the "Earth Summit" and brought together 172 nations and over 20,000 representatives of NGOs and other civil society actors to discuss the future of the global environment and "how we can reduce poverty, advance social equity and ensure environmental protection" under the banner of "sustainable development" (UN Brussels, 2011).

The summit's message was "that nothing less than a transformation of our attitudes and behaviour would bring about the necessary changes" to halt the destruction of irreplaceable natural resources and pollution of the planet (UN, 1997). The resulting documents laid the foundation for some of the most significant international attempts at confronting global environmental concerns, including the Rio Declaration on Environment and Development, the Statement of Forest Principles, and the Convention on Biological Diversity (CBD). The first COP to the Convention (COP1) convened in Berlin in March 1995 and began the groundwork for the Kyoto Protocol, adopted at COP3 in 1997. Since then, COP has been held annually in various cities around the world, including the two that are referred to in this book: Lima, 2014 (COP20) and Paris, 2015 (COP21). It is attended by environmental experts, ministers, heads of state, and NGOs, including IPOs.

In 2012, it was agreed at COP18 in Doha (despite significant difficulties in the negotiations) that only countries that had signed up for a second period of commitment to the Kyoto Protocol would be able to trade emissions allowances in the emerging carbon markets. By this time, market-based instruments (MBIs) had already become a fundamental part of United Nations (UN) climate change mitigation strategies (Pirard, 2012, p. 63). It was also agreed that negotiations for a new climate change agreement to replace the Kyoto Protocol would be concluded in time for COP21 in Paris. The same year, the UN held another major conference known as Rio+20,[8] which was intended to celebrate two decades since the Earth Summit[9] that had been held in the same city. Yet, in the months following Rio+20 environmental activists and organisations and Indigenous Peoples did not find cause to celebrate the event. Instead, criticism of the way in which the environment was being governed at the international level reached an all-time high. NGOs including the World Wildlife Fund (WWF) and Oxfam described the conference as nothing short of "disastrous" (Ford, 2012), and Greenpeace stated that the UN's position, which envisaged the development of a "green economy", was "devoid of meaning" (Tienhaara, 2014, pp. 193–194).

Many indigenous groups pre-empted the failure of Rio+20 and rejected it altogether, holding an alternate summit named "*Terra Livre*" (Free Earth). All 1,800 delegates from indigenous communities, Indigenous Peoples and IPOs including COICA, signed The Final Declaration of the 9th Terra Livre Camp, entitled *In Defense of the Commons: Against the Commodification of Life*,[10] denouncing REDD+ and carbon trading, which they described as "false solutions that do not solve environmental problems but seek to commodify nature and ignore the traditional knowledge and ancient wisdom of our peoples"[11] (Terra Livre, 2012b). The parallel "*Cumbre de los Pueblos*" (Peoples' Summit) also published a position paper, *Another Future is Possible*, criticising the UN vision that was outlined in *The Future We Want* (the official declaration adopted at Rio+20), and particularly its emphasis on carbon markets and the commodification of "ecosystem services" (Tienhaara, 2014, pp. 193–194).[12]

REDD+, PES, and the Commodification of Nature

REDD+ was conceived formally in 2007 at COP13 in Bali (Alexander et al., 2011, p. 683). Its precursor, REDD, was proposed two years earlier at COP11 in Montreal by representatives of Costa Rica and Papua New Guinea on behalf of the Coalition for Rainforest Nations[13] (CfRN) in a bid to "reconcile forest stewardship with economic development" (UN, 2015). Though it began as an attempt to combat *de*forestation, REDD became REDD+ to account for activities known as "carbon sequestration" that increase the removal of carbon from the atmosphere through *reforestation* and *afforestation* (Baldo-Soriano et al., 2012, p. 38). At the same time, the explicit aim of providing "co-benefits" to participating developing countries by providing opportunities for poverty alleviation and "sustainable development" was integrated into the REDD+ approach (Alexander et al., 2011, p. 683). The aims of REDD+ (along with a set of "safeguards"[14]) were officially agreed upon at COP16 in Cancún, 2010, and are described in paragraph 70 of the conference decisions as follows:

> [… REDD+ …] Encourages developing country Parties to contribute to mitigation actions in the forest sector by undertaking the following activities, as deemed appropriate by each Party and in accordance with their respective capabilities and national circumstances:
>
> a. Reducing emissions from deforestation
> b. Reducing emissions from forest degradation
> c. Conservation of forest carbon stocks
> d. Sustainable management of forests
> e. Enhancement of forest carbon stocks
>
> *(UNFCCC, 2010, p. 12)*

The way in which this is encouraged is through providing financial incentives to protect and expand forest stocks, which is generally measured in terms of tons of carbon dioxide equivalents (tCO_2e). REDD+ quickly became one of the most important and ubiquitous of all contemporary climate change mitigation strategies and has since been instrumental in shaping conservation policies in the Global South, carrying with it wide-reaching environmental, economic, and development implications (Corbera & Schroeder, 2011, pp. 89–90). Though projects vary from country to country, as they have materialised so too has the reality of REDD+: it is essentially a legal mechanism for incorporating forests into international financial flows, allowing stored carbon to be commodified and sold to (or through) the Global North in a bid to "offset" emissions (Dehm, 2012, p. 99). Part of the intention behind this approach was to begin to address the North-South "climate debt" under a polluter-pays logic that removes the market externality of pollution while facilitating the transition to a post-carbon society (Böhm, Murtola & Spoelstra, 2012, p. 7; Bond, 2012, p. 44). Ideologically, however, it was embedded within a broader of a broader shift towards a PES model that seeks to harness the power of the market in order to combat climate change.

The belief that the best way to protect the environment is to put a price on nature's so-called "services", to assign property rights, and to trade them on a global market has long been accepted across a wide range of countries and institutions (Liverman, 2004, p. 734). The projects and programmes that have emerged based upon this precept vary considerably, ranging from direct transactions between "providers" and "beneficiaries" of ecosystem services to broader schemes in which those who benefit from the services pay those who provide them indirectly (Cavelier & Grey, 2014, p. 3). This is usually facilitated through the sale and purchase of credits and offsets, which can be through carbon markets,[15] but it can also include a wide range of financial mechanisms such as green or blue bonds, biodiversity offsets, and biodiversity banking.[16] What is consistent across all approaches, though, is that they are based upon the theoretical assumption that the undersupply of ecosystem services[17] is the result of market failures, and so accounting for these by creating a system through which they are valued (in the economic sense) and paid for (in the monetary sense) can remove these "externalities". This also rests on the assumption that because providers of ecosystem services can often be found in rural, indigenous, and other marginalised communities in the Global South, PES can, indeed *will*, contribute to poverty alleviation and development through "win-win" scenarios (Arsel et al., 2013, p. 274).

There are many concerns associated with this idea, but Corbera and Kosoy (2010, p. 1228) offered an early summary those that are most relevant to this book: first, "narrowing down the complexity of ecosystems to a single service has serious technical difficulties and ethical implications on the way we relate to and perceive nature", second, "the commodification of ecosystem services denies the multiplicity of values which can be attributed to these services, since it requires that a single exchange-value is adopted for trading", and finally, "the process of production, exchange and consumption of ecosystem services is characterised by power asymmetries which may contribute to reproducing rather than addressing existing inequalities in the access to natural resources and services". More broadly, the commodification of nature that REDD+ and PES facilitate represents an attempt to "compel nature to pay for itself", which has been described by Büscher and Fletcher (2015, p. 273), as a process of Accumulation by Conservation (AbC). With echoes of David Harvey's theory of accumulation by dispossession (2011, pp. 48–49), Büscher and Fletcher argue that

> AbC is an effort to obfuscate the daunting implications of capitalist production by claiming that capitalism has the ability to effectively address these problems through the same mechanisms that created them. In this sense, AbC can be viewed as something of a "pre-emptive strike" precluding any possible chance for the development of sane, animated nature–society engagements.
>
> *(Büscher & Fletcher, 2015, p. 293)*

So, despite declarations in various documents (such as the 2011 United Nations Development Programme Human Development Report [UNDPHDR, 2011] and the 2008 United Nations Declaration on the Rights of Indigenous Peoples [UNDRIP, 2008]) that a truly sustainable form of development will involve a *human* change, rather than an *economic* one, and the message from the 1992 Rio Summit highlighting the need for "nothing less than a transformation of our attitudes", the UN has so far unwaveringly adhered to neoliberal economic principles[18] of development. As such, the same market mechanisms that have contributed to both the inequity and unsustainability[19] that we see today continue to be treated as the primary remedy for the problem. The projects that have emerged under the auspice of PES and REDD+ are indicative of a more general trend of non-indigenous people and institutions coercing Indigenous Peoples and their communities into market systems (Cabello & Gilbertson, 2012, p. 163) and accumulating territories by dispossessing local people (Harvey, 2011, pp. 48–49).

COICA: Indigenous Alternatives to REDD+

On a Wednesday morning in January 2015, a month after returning to Quito from COP20 in Lima, I stood at the door to the head office of COICA. I arranged the visit beforehand, but had been trying to do research with Indigenous people in Ecuador for almost six months, and I had become accustomed to people agreeing to meet with me and then changing their mind at the last minute or just not showing up. I half expected to be turned away, but I hoped to at least interview a member of the organisation. I rang the bell, holding it down for a while to make sure it was heard. I waited, and waited, knocked on the door, called the office (no answer), and eventually gave up and walked home. Later, I called again and asked what had happened. I was told everyone had been out of the office, but I was invited to return the next day. I did, and to my surprise, I was welcomed warmly by the manager of the office, Tomás. I guessed he was older than me, but not by much. He was smartly dressed in a pressed pink and white pin-striped shirt, black trousers, and black leather shoes. To me, he appeared *Indigenous* despite his clothing because of his dark skin and hair, but he was different to the other Indigenous people I had spent time with who had mostly dressed either casually or "traditionally". He was both professional *and* Indigenous. We went to his office and had a very brief conversation. He asked how he could help:

> "My name is George Byrne, I am from the University of Sussex in England and I am doing research into indigenous alternatives to REDD+" I stumbled over my broken Spanish.
>
> "*¡Bueno!* I read your email. And you want to work with us?" Tomás asked.
>
> "Well, I am going to be here in Ecuador for nine more months, *más o menos*, and can help out in the office if you like". I wasn't expecting it to be this easy.
>
> "OK, *como una pasantía*" I didn't understand what he said, but he didn't wait for a response "do you need a desk? It is very busy here today, so do you want to start on Monday?"
>
> I wasn't entirely sure what he meant by "start", either, but decided to figure that out later.
>
> "Yes, thank you, if it's possible?" I responded.
>
> He laughed and said "Good, I need to practice my English!"

Tomás then stood up and gestured for me to follow him. He took me to a small reception desk and introduced me to the woman behind it but did not tell me her name.

> This is … Jorge? he is from England and is an expert on REDD+, he is going to do a *pasantía* with us for nine months, can you give him the Wi-Fi password. Jorge, when you come in on Monday you can have one of those desks up there.

He pointed to a mezzanine level, up a small flight of stairs, where there were three desks. Two of them were empty, and one looked like it was in regular use, though nobody was there. Tomás said goodbye and went back to his office, leaving me to get the Wi-Fi password. I thanked the receptionist and left, feeling surprised and happy. When I got outside, I looked up *pasantía* on my phone: "internship", said Google Translate. I thought that was what he said, and I remembered that it had been mentioned as a possibility in an email, and though I did not really know what it would involve, I felt I should help out if I could, rather than just asking for their time. My limited experience of attempting to work with Indigenous people in Ecuador had so far been difficult and had left me with low expectations, assuming nothing would go smoothly. But it seemed that for COICA, working with a white man who turns up at the door was viewed more positively than it had been in the indigenous community and organisations I had previously visited. The office also felt familiar to me, like it could be the office of any NGO in any other city, and it also felt safe.

Once outside the office, I became aware of how hot it was. It would have only been 20-something degrees, but the midday heat in Quito is oppressive. The altitude and position of the city, which is almost on the equator, means that it sometimes feels as though the sun is barely out of arms reach, and I burn very easily. I could either pay a couple of dollars for a taxi home, or I could walk the kilometre or two down to *El Camino de Orellana*, where I lived with my wife on the mountain side. Despite the heat, and despite knowing it is not a great idea to wander around with a laptop, a camera, and a phone in this part of Quito, I started walking. A few months earlier, my wife and I had been attacked, kidnapped, and robbed in what had appeared to be a taxi like any other, and since then I had walked everywhere in the city at any time of day or night, even through dark and dangerous neighbourhoods. It felt safer than getting into a car with a stranger because I knew I could run.

As I walked home from the COICA office, I turned a corner and saw three young men walking towards me. Like the taxi, there was nothing particularly unusual about them, other than the way they had spread out across the street. As we passed each other, I felt surrounded by them, and I panicked. I don't remember running home, but I know that I did because I remember stopping outside my house and vomiting in the gutter before falling to the floor. I sat there - I have no idea for how long - dizzy, sweating, and trying to calm myself. It was not the first panic attack I'd had recently, and it wouldn't be the last, but I had not yet found a way to make sense of or manage such irrational responses to fear. The thing I was most afraid of at that time was that I did not recognise myself – this wasn't me.

The following morning, Tomás called and asked if I could come in on the weekend and help with an English test. I didn't quite understand what he wanted me to do, but I agreed to meet him at the office, eager to begin spending time there and happy to be helpful if I could. He had mentioned before that he needed to practise his English, so I assumed he had an exam coming up and wanted help preparing for it, but that was not it at all. I left my house early so that I could walk the long way round and arrived at nine on Saturday morning, as agreed. Tomás answered the door with a smile and led me to his office, passing the main meeting room on the way. The room had been empty last time I was there, but this time there were approximately 15 smartly dressed women sitting quietly on chairs placed around the edges of the room. Once in his office, Tomás told me (in Spanish) that he wanted me to go into the room and explain (in English) the requirements of the exam the women were about to take. He then wanted me to mark the exams and rank the scores of the women, who he told me were applying for a job as a receptionist. We walked into the room, he introduced me as "Professor Jorge" from England, and nervously I did as he had asked. I started speaking in Spanish, then remembered I was supposed to be speaking in English, which felt strange. It was odd to be put in this position, and it was also uncomfortable to have been introduced to strangers as both an "expert" and a "professor" in the space of a few days, neither of which I had ever claimed to be.

*

In 1984, representatives of Amazonian indigenous federations from three countries (Interethnic Association for the Development of the Peruvian Forest [AIDESEP] from Peru, Confederation of Indigenous Nationalities of the Ecuadorian Amazon [CONFENIAE] of Ecuador, and the Confederation of Indigenous Peoples of the Bolivian Orient [CIDOB] of Bolivia) and two national IPOs (Organization of Indigenous Nationalities of Colombia [ONIC] and Union of Indigenous Nations of Brazil [UNI]) came together to form COICA, which would act as an umbrella organisation to advocate for Indigenous Peoples at the international level (Climate Alliance, n.d.). It was a strategic alliance set up in preparation for participation in the UN Working Group on Indigenous Populations, which met in Geneva the same year. Thus, from its very beginning, the organisation was designed to work with and within the development institutions of the Global North. In 1993, COICA established a head office in Quito, Ecuador, and, by 1994, had expanded to include the Indian National Council of Venezuela (CONIVE), the Association of Amerindian Peoples of Guyana (APA), the Organization of Indian Surinam (OIS), and the Federation of Indigenous Organizations of Guyana (FOAG) making it the first and only coordinator and representative of Indigenous Peoples' federations from all nine Amazonian countries.[20] This and the organisation's success in rapidly building alliances with human rights NGOs, environmental organisations, and multilateral agencies meant that COICA quickly became, at the diplomatic level at least, the most prominent representative of Amazonian Indigenous Peoples (Mato, 2000, pp. 353–354). In the late 1980s and early 1990s, COICA was particularly successful at building partnerships with municipalities in European cities that were seeking ways to combat the greenhouse effect and deforestation, leading to the formation of the Climate Alliance of European Cities with Indigenous Rainforest Peoples in 1990. The Climate Alliance then expanded considerably, becoming a network through which

> 1,700 member municipalities and districts covering 26 European countries as well as a variety of regional governments, NGOs and other organisations are actively working to combat climate change.
>
> *(Climate Alliance, n.d.)*

Despite the expansion of the European side of the Climate Alliance, COICA remained its sole indigenous partner organisation. In the United States, COICA also developed relationships with numerous NGOs, eventually leading to the first Amazon Initiative Conference of NGOs and COICA in 1993, and subsequently the formation of the "Amazon Alliance", which acted as a coordinating body composed of IPOs and NGOs with a permanent coordinating office in Washington (Pieck, 2013). Though the Amazon Alliance no longer exists, many of the key organisations involved in it have continued to work directly with COICA.[21] These long-term strategic alliances have enabled COICA to access substantial funds through international cooperation, but in addition to this, its visibility and alliances have allowed the organisation to secure a unique presence in Intergovernmental Organizations (IGOs) where Indigenous Peoples, international development, and the environment are discussed, particularly the World Bank and the UN. As such, COICA has been the primary Amazonian indigenous presence at COP conferences and in the negotiations that have shaped policy documents such as the Draft UNDRIP, the CBD, and the UNFCCC.

Despite its cooperation with the UN, COICA was initially sceptical of the REDD+ approach. In 2008, in collaboration with The Nature Conservancy (TNC), the German Technical Cooperation Agency (GTZ), the WWF, and the Norwegian Agency for Development Cooperation (NORAD), COICA produced a pamphlet entitled *REDD and Its Implications*, which was intended as a manual to be distributed to Indigenous Peoples explaining key concepts and concerns surrounding REDD+. In the document, two important observations were made that would contribute to the eventual emergence of REDD+ Indígena Amazónico (RIA). First, it listed the measurement of carbon (along with market mechanisms, monitoring, distribution of benefits, risks of corruption, and implications for land rights) as a principal problem of the REDD+ approach. Second, it noted that

> For Indigenous Peoples, protecting the forest and avoiding soil degradation has been a practice for millennia. With REDD or without REDD, Indigenous Peoples are protecting the forests and biological diversity of the Amazon [author's translation].
>
> *(COICA, 2008, p. 14)*

Which was followed by a statement that summarised why, if Indigenous Peoples would be protecting forests anyway, they were so concerned about REDD+:

> What worries the indigenous organisations is that in the name of REDD, States violate Indigenous Peoples' collective rights. It is of great concern that they are not given a proper space in the decision-making process of establishing the REDD mechanism and its implementation [author's translation].
>
> *(COICA, 2008, p. 15)*

RIA was developed in response to these concerns. It is different to what COICA refers to as "conventional" REDD+, in that it advocates for a holistic approach to forest conservation and is based upon developing tailored "planes de vida plena" (broad life plans) for each indigenous community where a RIA project is to be implemented. It also intends to go "más allá" (beyond) carbon (COICA, 2013, pp. 2–3), and "values the integrity and interconnection of ecosystem services" (COICA, 2015, p. 5).

I first encountered COICA at COP20 in Lima, in 2014, where the organisation, in collaboration with the Government of Peru, had been instrumental in establishing the first Indigenous Peoples Pavilion. The Pavilion was an exhibition space and auditorium with thematic panels for Indigenous Peoples and a parallel space in the public Voices for Climate area. The Government of Peru's Ministry of the Environment (MINAM) described it as facilitating:

> ... processes and spaces in the framework of the UNFCCC to enable the indigenous peoples of the world, [...] based on their worldviews and principles of management of their Territories, to express their insights on climate crisis, develop and present their proposals and alternatives and to participate in the specific negotiations of the Conference.
>
> *(MINAM, 2014)*

At the Pavilion on December 9, COICA announced the launch of two pilot RIA projects in Madre de Dios (Peru) and Inirida (Colombia) as well as a "test case" that would be implemented in Ecuador the following year. The aim of the projects, supported through WWF Germany and the Government of Germany, was to begin integrating elements of RIA into regional and national climate change strategies (WWF, 2014). Though it is not the intention of this book to discuss the strengths and weaknesses of RIA, it is important to note that to successfully align the project with national programmes and access the significant financial resources mentioned above, COICA continues to use the discourses associated with PES and carbon trading. In doing so, the "holistic" approach has been translated by COICA into 24 "functions" (as opposed to "services"), which is at times illustrated using maps that again reduce the argument further by demonstrating that forest carbon stocks are safer and more abundant in areas where Indigenous Peoples live and have land rights.

RIA, then, depends upon a shared belief that the holistic approaches of Indigenous Peoples allow them to be stewards of the forest, and yet proving this to a non-indigenous audience, especially within the context of the UN climate negotiations, means measuring this contribution in terms of functions and tCO_2e.

By bridging this gap, COICA has been instrumental in advancing the notion of stewardship in environmental discourse within the UN and more broadly. In 1990, before the Earth Summit, and even before the UNFCCC, COICA became highly critical of emerging "debt-for-nature" swaps (whereby portions of a developing nations debt could be "swapped" for commitments to national environmental conservation measures). This was because it put organisations (particularly the WWF who put forward the model) in

> ... a position of negotiating with our governments the future of our homelands [...] We know of specific examples of such swaps which have shown brazen disregard for the rights of indigenous inhabitants [...] we propose joining hands with those members of the worldwide environmentalist community who recognize our historical role as caretakers of the Amazon Basin.
>
> *(COICA, 1990, cited in Alston & Brown, 1993, p. 189)*

In response, COICA offered an alternative to debt-for-nature deals, which it called "debt-for-Indian-stewardship" swaps, through which Indigenous Peoples who have an ancestral claim to land would be able to benefit from them and maintain access to them (Alston & Brown, 1993, pp. 189–190). In a similar way RIA harnesses the idea of indigenous "ecological nobility", or stewardship, in a bid to secure territorial rights while simultaneously gaining direct access to the growing international funds being allocated to protecting rainforests through REDD+. The first two of the four ethnographic chapters (Chapters 3 and 4) discuss in much more detail the complex role of COICA in shaping indigeneity, including at the Indigenous Pavilion at COP.

The Sápara: Indigenous Resistance to REDD+

Before encountering COICA at COP20, I began my fieldwork in a small indigenous community in the Amazon region of Ecuador. I conducted two short trips with the intention of eventually extending this to a long-term ethnographic study into how the people who lived there understood and interacted with the emerging national REDD+ readiness programme called "Socio Bosque". Following the drafting of the new Ecuadorian Constitution in 2008, which stressed the need to face climate change and to reduce deforestation, the Ecuadorian Ministry of the Environment (MAE) designed Socio Bosque as a voluntary initiative offering economic incentives for forest conservation (Nuñez, 2011, p. 9). Though first conceived separately from REDD+, Socio Bosque shared many of its principles and therefore provided a suitable platform from which to launch Ecuador's REDD+ readiness programme (Bertzky et al., 2010, pp. 1–3). It became a key part of the Government of Ecuador's (GOE's) environmental strategy, along with parallel initiatives focussing on forest information, forest control, sustainable forest management, reforestation/afforestation, and land ownership (Hübenthal, Starnfeld, & Carrión, 2010).

Socio Bosque is underpinned by four principles: to be simple and transparent, legally enforceable, fair and equitable, and non-prohibitive for participants. Its approach is quite straightforward, offering set payments per hectare of maintained forest cover, differentiated only by the size of the area under contract. Voluntary 20-year contracts are subject to regular checks through satellite imaging, aerial photography, and field visits that include social monitoring and reviews of spending reports against investment plans. Annual payments are dependent on strict compliance and demonstrable outcomes (Fehse, 2012, p. 1). While the GOE has funded and facilitated major initiatives through Socio Bosque, additional financial resources are crucial to sustaining the Programme, and it is to this end that Socio Bosque was positioned within the broader framework of REDD+ (Lawson, Maginnis, & Suarez, 2010, pp. 47–48).

In Ecuador, Indigenous Peoples' interactions with the Government and other actors who have taken an interest in forest lands and resources have historically been tense, and both REDD+ and Socio Bosque have also at times been resisted. Some IPOs and environmental NGOs working in

Ecuador have criticised the UN approach to financing conservation, describing it as a "new face for capitalism" that extols "the paradigm that the solution lies in the market, in property rights, in [...] the commodification of all of nature" (Acción Ecológica, 2012). As the UN acknowledges, that forest conservation measures may indirectly force displacement of indigenous groups (Barnsley, 2009, pp. 36–37), and given the disastrous impacts of past encroachments on indigenous lands in Ecuador (see, e.g., Gerlach, 2003; Sawyer, 2004), indigenous resistance to the implementation of policies pertaining to Ecuador's forests is somewhat unsurprising. Moreover, the general criticisms of the REDD+ approach began to be realised in the experiences of marginalised peoples in participating countries, including Ecuador (see, e.g., Arts et al., 2018; Corson & MacDonald, 2013, pp. 37–38), which intensified these fears.

In light of these concerns, CONFENIAE (the umbrella organisation of Amazonian Indigenous Peoples in Ecuador, which is also part of COICA) initially rejected REDD+ negotiations outright stating that the project intended to "take away our freedom to manage our resources" (Lang, 2009), and the Confederation of Indigenous Nationalities of Ecuador (CONAIE) (the national IPO that also represents Indigenous Peoples who reside outside of the Amazon) explicitly rejected the implementation of REDD+ and Socio Bosque claiming that the programmes seek to privatise indigenous lands describing REDD+ as "perverse" and arguing that it "is not a solution to climate change [it] has been created by multilateral institutions like the World Bank that routinely violate Indigenous Peoples' rights and pollute Mother Earth" (Boas, 2011, pp. 46 and 83; Reed, 2011, p. 535). In 2012, the MAE also acknowledged the shortcomings of its Socio Bosque projects, particularly regarding community participation, citing a report by the REDD+ Social and Environmental Standards Organisation, which found that

> The communities visited [... in Ecuador ...] professed to not having taken part in processes of prior consultation or free, prior and informed consent. The decision to join Socio Bosque was communicated in general assemblies, which can exclude marginalized groups such as women, youth and seniors. Even if information is received by community leadership, this does not ensure that effective consent has been granted, nor that the information is available to all members of the community [...] Though prohibited by the Constitution, cases of physical displacement by different causes were observed.
>
> *(MAE, 2012a, p. 1)*

Despite these failings and the tense relations they engendered, Socio Bosque proposals were well received by leaders of some indigenous communities and were not perceived as threat. For example, José Quenamá, a leader of the Cofán people, stated "we're happy with [Socio Bosque], because we have an income and they have told us we can keep hunting, that we can use the wood we need. We don't know anything more than that" (Saavedra, 2011). Other communities feared that ambiguities in the Socio Bosque programme regarding sanctions for non-compliance would leave indigenous land rights and the right to use forest resources in a precarious position. Paco Chuji, president of the Federation of Organizations of Kichwa Nationality of Sucumbíos-Ecuador (FONAKISE), worried that

> When communities are critical of the government or do not go along with its actions, the felling of trees to build houses in the community may be interpreted as a breach of the agreement, and they will receive sanctions that nobody knows precisely the nature of.
>
> *(Saavedra, 2011)*

There are also contradictions regarding the relationship between Socio Bosque and the extractivist industries in Ecuador. A concern is that land supposedly protected by Socio Bosque (due to the constitutional framework and national laws that give the government the right to control "strategic" resources) could still be subject to oil and gas extraction, even if the use of the forest by indigenous communities for their livelihoods is restricted. In 2012, The Indian Law Resources Center wrote to the co-chairs of the UN-REDD Programme Policy Board requesting an investigation into "the impact of the oil and gas concession process on Socio Bosque, on the rights of Indigenous peoples, and on the objectives of the UN-REDD Programme" (ILRC, 2012, p. 2). The letter also requested that the Policy Board identify concrete actions to be taken by UN-REDD and the GOE to protect the human rights of Indigenous Peoples, and it was accompanied by letters to the GOE from CONAIE, the Sápara Nation, and other indigenous groups in Ecuador highlighting the detrimental role of ongoing oil extraction (Lang, 2012). But even though Sápara leaders and many other Amazon Indigenous Peoples have explicitly rejected all oil and gas industry activities on their lands, the bidding process has continued, including in the oil concession blocks that are located close to the community I visited.

In 2009, the Sápara Nation signed a contract with the MAE agreeing to participate in the GOE's Socio Bosque programme. According to the UN Sustainable Development Solutions Network, this meant 121,682 hectares of Sápara land would now be "conserved" under the programme and there would be 3910 "beneficiaries" of the project (SDSN, 2015, p. 72). The MAE states that improving "the living conditions of the owners who are responsible for the conservation of the forests and paramos[22] and to guarantee that the resources delivered are used in a transparent manner based on the internal consensus of the beneficiaries" is a key objective of Socio Bosque (MAE, 2012b). However, the Sápara people in Ecuador number 400 at most, and in the community I visited (where over 40 Sápara people live), residents claimed that they had not benefitted from Socio Bosque at all, which raises the question of who, precisely, these thousands of "beneficiaries" are. An initial assessment by the Indigenous Peoples' Biocultural Climate Change Assessment Initiative (IPCCA) found that the MAE, under the rubric of Socio Bosque, had mapped Sápara territory using satellite imaging to detect and prohibit new settlements or *chakras* (small communal gardens) inside the area. The IPCCA report also claimed that this was done without the free, prior, and informed consent of the Sápara people and that "From the point of view of the communities, it is counterproductive that they are not allowed to continue occupying and managing their territory in the traditional way. They feel that *they are being demoted to mere guardians of their own forests* [emphasis added] and, in practical terms, are no longer owners of their land" (IPCCA, 2011, p. 7).

During my research, I learned that the community I visited was not located within the area that is included in the Socio Bosque project, which could explain why they had not received any financial or other benefits from Socio Bosque payments and why oil extraction remains a possibility. However, it was explained to me during the workshops that I conducted on behalf of the IPCCA that the community had believed the benefits would be distributed equitably across the Sápara Nation, regardless of whether individuals lived within the designated Socio Bosque area. The President of the Sápara Nation in Ecuador at the time, Kléver Ruíz, confirmed to me that little or nothing had been received by the Sápara Nation as a whole, and that no democratic process was undertaken to ensure that the Sápara people had given their consent. Within the community, there were also accusations that when signed consent was supposedly collected from the potential beneficiaries, many of the signatures were forged by government officials who visited communities and *consulted* them but did not gain their *consent*. There was even some disagreement about whether the contract had ever been signed by a legitimate representative of the Sápara people, or if the document itself had been forged. It is unsurprising, then, that Socio Bosque, and by extension REDD+, was generally seen as a way for the government and oil companies to gain control of their land, which was seen as part of a bigger plan to gain access to oil and other resources in their territory. The most common response I received when I asked members of the Sápara community what they thought about Socio Bosque and REDD+ was "It is a lie!".

Conclusion

This chapter is intended to provide some insight into why my research began where and when it did, and why I chose to focus on questions relating to Indigenous Peoples and REDD+ in Ecuador. It should be noted, however, that the historical relationship between the various actors mentioned (Indigenous Peoples, IPOs, NGOs, Governments, and IGOs) is very complex, and due to the constraints of this book, it is only possible to provide a cursory overview. What is also important to note is that even in this chapter there are some allusions to the processes of selection that will be mentioned throughout the book. For example, not all Indigenous people in Ecuador (or elsewhere) have had negative experiences of REDD+ and other PES projects, but from early in my fieldwork, I began searching for those who had been critical of such approaches. This was based upon my pre-existing bias *against* the use of MBIs in attempting to combat climate change. I attempted to work with the Sápara in part because I knew they had a complicated relationship with these projects, as opposed to the Cofán (for example) who had so far had fairly positive experiences of Socio Bosque. Also, at the international level, it is apparent that particular Indigenous people, especially COICA, have been very successful in securing a position in international negotiations that pertain to climate change and development. Though they have been critical of PES, COICA has also sought to work within or alongside the UNFCCC framework and has chosen to adopt, rather than reject, some elements of the discourse of REDD+ in the form of RIA. In doing so they have propagated certain ideas regarding indigenous "ecological nobility" and "stewardship" as well as translating "holistic" approaches to forests into the language of "functions" and their relation to carbon sequestration in units of tCO_2e.

The chapters later in this book illustrate how particular notions of indigeneity emerge in the context of sustainable development, and how asymmetries in access to resources and information, and inequitable access to and representation in development institutions shape what it means to be *Indigenous* in relation to climate change. First, however, I discuss some of the methodological complications that arose when considering this issue, not least that this book is itself embedded within, as well as being implicated in upholding, the same inequitable structures of power that I am critiquing.

Notes

1 For discussion of the emergence of two parallel research threads in relation to the "ecologically noble savage", the first of which focuses on resource use among indigenous communities, while the second considers ecological nobility

in terms of identity and knowledge, as well as its use as a political tool, see Hames (2007).

2 Hereafter, the "Indigenous Pavilion".

3 The RCP is described as "a platform for raising awareness and sharing information about the latest practices and scientific findings linking biodiversity, climate change, and sustainable land management" (RCP, n.d.). It was established following the Earth Summit in 1992, during which the three Rio Conventions were established: The Convention on Biological Diversity (CBD), the United Nations Framework Convention on Climate Change (UNFCCC), and the United Nations Convention to Combat Desertification (UNCCD). This RCP was convened at COP21 for the tenth time since being established in 2010 and convened alongside CBD COP10, held October 19–29, 2010, in Nagoya, Japan. At COP21 in Paris, thematic days included biodiversity and ecosystems; local communities and Indigenous Peoples; land; oceans; implementing the Paris Agreement; and gender (IISD, 2016, p. 1).

4 The panel also included three more experts on REDD+: Pavan Sukhdev (CEO of GIST Advisory and UNEP Goodwill Ambassador), Maria Kiwanuka (Senior Presidential Advisor and former Finance Minister of Uganda), and Ivannia Quesada Villalobos (Vice Minister of Agriculture and Livestock in Costa Rica).

5 The new UN-supported Sustainable Stock Exchanges (SSE) Initiative is intended to facilitate the transformation of stock markets into instruments of "sustainable development" and "green growth", serving "as a tool for sustainable development, but also as a promoter and facilitator of new international investment opportunities, specifically for international and institutional investors in their drive to enlarge and diversify their portfolio" (Klagge & Zademach, 2018, p. 92). Marien and Sales position the SSE within a broader process of "greening capitalism, quietly" that they argue amounts to a process of "improvement, but growing inadequacy" (2017, p. 164) and cite Dauvergne who has argued that "most multinational corporations have become more proactive in managing critics, avoiding obvious greenwash, and instead partnering with NGOs, offering eco-products and sponsoring third-party certification of production processes and consumer products...(which) can make it seem as if rapid progress is now being made toward global sustainability. However, the efficiency gains of ecobusiness are largely lost as firms reinvest energy and cost savings to stimulate even more unsustainable growth and consumption – a rebound effect that's at the heart of the failure of environmentalism of the rich to slow the escalating global sustainability crisis" (Dauvergne, 2016, pp. 11–12).

6 The Indigenous Environmental Network has been particularly vocal in criticising REDD+ and has established a Global Alliance Against REDD+ which recently published a report along with the Climate Justice Alliance (made up of 58 urban and rural frontline communities, organisations and supporting networks in the climate justice movement) entitled *Carbon Pricing: A Critical Perspective for Community Resistance*. The report states that REDD+ projects tend to follow a divide-and-rule strategy. Despite promises of compensation for not using their forests, or even for simply continuing their customary practices without increasing deforestation. In reality "communities often find themselves subject to new restrictions on their livelihood activities, new accounting burdens, and even overt land grabs and criminalization, while the promised money is often not forthcoming and internal community tensions increase" (Gilbertson, 2017, p. 29).

7 A recent report by Friends of the Earth argues that the neoliberal logic of offsetting and carbon trading has meant that REDD+ initiatives fail to reduce

emissions, lead to violations of environmental and human rights, and legitimise inequality, while fuelling conflict within communities and undermining the autonomy of Indigenous Peoples to govern their own territories (Furtado, 2018). In 2017, 13 NGOs including The Rainforest Foundation and The Climate Alliance criticised the World Bank's Forest Carbon Partnership Facility (FCPF), which is a major funding stream for REDD+ readiness, on its tenth anniversary in an open letter that stated: "deforestation assessments and associated actions target the livelihoods of the poorest while baseline reference scenarios have been set so high in some cases as to allow for deforestation to increase but REDD "credits" still to be generated. The impacts of this lack of "REDD-readiness" are already being felt on the ground with worrying reports of land grabbing, conflict over carbon rights, and human rights abuses", describing the approach as "a monstrously inefficient and cumbersome way to make financial interventions to protect forests" (Rainforest Foundation UK, 2017, pp. 1–3).

8 Rio+20, officially "The UN Conference on Sustainable Development" (UNCSD) was organised with the aim of fulfilling General Assembly Resolution 64/236, which, in 2010, decided that a UN Conference on Sustainable Development would be held in 2012. Brazil offered to host the conference 20 years after hosting the landmark Earth Summit (UNCSD, 2012).

9 The 1992 Earth Summit, officially the "UN Conference on Environment and Development" (UNCED) was initially seen as a success and did make major promises regarding "sustainable development", though many critics argued that it did not go far enough (Batterham, 2003, p. 2167).

10 Original Portuguese *Em Defesa dos Bens Comuns, Contra a Mercantilização da Dida.*

11 Unofficial translation into English by Earth Peoples. Original Portuguese: "falsas soluções que não resolvem os problemas ambientais e procuram mercantilizar a natureza e ignoram os conhecimentos tradicionais e a sabedoria milenar de nossos povos". (Terra Livre, 2012a).

12 The notion of "the commons" and the related problem of commodification of them are themes that persistently arise in discussions about how (and why) the global environment will be governed. For a summary of "global commons", and the set of global environmental issues associated with the misuse of these unattainable or unprofitable resources (those that affect the atmosphere, the climate, the high seas and Antarctica), see O'Neill (2009, pp. 29–38) and Schrijver (2016).

13 The CfRN is an intergovernmental organisation of over 50 tropical countries, most of which would be described as "developing" or "under developed", but its headquarters are in New York. Since proposing REDD in 2005, capacity building (which includes readiness, funding, measurement, reporting, and verification) for national REDD+ programmes has been its main activity (see CfRN, n.d).

14 The Cancún safeguards were intended to ensure that projects implemented by governments and other developers achieve better protection for local communities. But their effectiveness in achieving this goal has been questioned, in part because the safeguards "employ ideas of development involving transformation of local communities (e.g. poverty reduction and clarifying issues related to tenure rights). This suggests a confusion and conflation of different objectives embodied in the concept of safeguards which could have dire implications" and as such "The different understandings of what constitutes safeguards can lead to a variety of interpretations of the term which has a range of implications for REDD+ policy and practice" (Arhin, 2014, p. 2).

15 Despite multiple falters, and doubts about their effectiveness and stability, the World Bank claims carbon markets are "back from the brink of collapse"

(Harvey, 2018). Ironically, this makes them one of the few endangered species that have been successfully saved from extinction in recent years.

16 Spash (2015, p. 541) argues that biodiversity banking and offsets, like other PES systems, use "economic logic to legitimise, rather than prevent, ongoing habitat destruction [...and...] operationalise trade-offs that are in the best interests of developers and make false claims to adding productive new economic activity".

17 According to the UNDP, ecosystem services come in four main categories "1. Provisioning services (the products obtained from ecosystems such as food and fresh water); 2. Regulating services (the benefits obtained from the regulation of ecosystem processes such as air quality and pollination); 3. Cultural services (the non-material benefits that people obtain such as spiritual enrichment, recreation and aesthetic experiences) that directly affect people; and 4. The supporting services needed to maintain the other services (such as photosynthesis and nutrient recycling)" (UNDP, 2015).

18 In recent years, some significant actors at the UN have openly criticised neoliberalism, including when the United Nations Conference on Trade and Development (UNCTAD) chief Mukhisa Kituyi stated in a news conference in 2017 that "The whole neo-liberal mantra that 'there is no alternative' has begun to fall apart" and that there are, in fact, "plenty of alternatives out there and they are urgently needed given the kind of economic and social imbalances that we are currently facing" (Miles, 2017). Though UNCTAD, which I had imagined would pro-market, has some history of anti-capitalist sentiment and acknowledges that cracks are forming in neoliberal approaches to development, market-based mechanisms, and particularly carbon pricing and remain deeply embedded in the UNFCCC framework, including in the Paris Agreement (see, e.g., Stavins & Stowe, 2017).

19 See, e.g., Bellamy Foster (2012) and Harvey (2011).

20 The Interethnic Association for the Development of the Peruvian Forest (AIDESEP), the Confederation of Indigenous Nationalities of the Ecuadorian Amazon (CONFENIAE), The Confederation of Indigenous Peoples of the Bolivian Orient (CIDOB), the Organization of Indigenous Nationalities of Colombia (ONIC), now known as the National Organisation of Indigenous Peoples of the Colombian Amazon (OPIAC), the Union of Indigenous Nations of Brazil (UNI), now known as the Coordination of Indigenous Organizations of the Brazilian Amazon (COIAB),the Indian National Council of Venezuela (CONIVE) now the Venezuelan Regional Organisation of Indigenous Peoples of the Amazon (ORPIA), the Association of Amerindian Peoples of Guyana (APA), the Organization of Indian Suriname (OIS), the Federation of Indigenous Organizations of Guyana (FOAG).

21 Including the Environmental Defense Fund (EDF), Oxfam, the Rainforest Action Network (RAN), and the World Wildlife Fund (WWF).

22 Páramos are Andean grasslands, and Ecuador has a programme that runs alongside Socio Bosque which provides a similar PES incentive for their conservation "Socio Páramo" (Bremer, Farley & Lopez-Carr, 2014, p. 122).

References

Acción Ecológica. (2012). *Open letter to the Secretariat of Convention on Biological Diversity and the Governments of Ecuador, India, Japan, Norway and Sweden.* Retrieved from https://climateandcapitalism.com/2012/02/12/un-biodiversity-plan-neoliberal-measures-to-commodify-nature/

Alexander, S., Nelson, C. R., Aronson, J., Lamb, D., Cliquet, A., Erwin, K. L., & Murcia, C. (2011). Opportunities and challenges for ecological restoration within REDD+. *Restoration Ecology*, *19*(6), 683–689.

Alston, D., & Brown, N. (1993). Global threats to people of colour. In R. D. Bullard (Ed.), *Confronting environmental racism: Voices from the grassroots* (pp. 179–194). Boston, MA: South End Press.

Arhin, A. A. (2014). Safeguards and dangerguards: A framework for unpacking the black box of safeguards for REDD+. *Forest Policy and Economics*, *45*, 24–31.

Arsel, M., Adaman, F., Agarwal, B., Aguilar, B., Corbera, E., Ezzine de Blas, D., ... Pellegrini, L. (2013). Payments for ecosystem services and the fatal attraction of win-win solutions. *Conservation Letters*, *6*(4), 274–279.

Arts, B., Carter, S., Gillerd, K. E., Golcherc, S. C., Herold, M., Kok, K., ... Wollenbergj, E. (2018). Climate-smart land use requires local solutions, transdisciplinary research, policy coherence and transparency. *Carbon Management*, *9*(3), 291–301.

Baldo-Soriano, E., Carling, J., Chavez, R., Erni, C., Martone, F., & Tugendhat, H. (2012) *What is REDD+? A guide for indigenous communities* (3rd ed.). Retrieved from http://www.forestpeoples.org/sites/fpp/files/publication/2012/09/what-redd3rd-edition-september-2012.pdf

Barnsley, I. (2009). *Reducing emissions from deforestation and forest degradation in developing countries (REDD): A guide for indigenous peoples*. Yokohama: UNU-IAS Guide.

Batterham, R. J. (2003). Ten years of sustainability: Where do we go from here? *Chemical Engineering Science*, *58*, 2167–2179.

Bellamy Foster, J. (2012). *The four laws of ecology and the four anti-ecological laws of capitalism*. Retrieved from http://climateandcapitalism.com/2012/04/02/four-laws/

Bertzky, M., Ravilious, C., Araujo-Navas, A. L., Kapos, V., Carrión, D., Chiu, M., & Dickson, B. (2010). *Carbon, biodiversity and ecosystem services: Exploring co-benefits. Ecuador*. Retrieved from http://www.unep.org/pdf/pressreleases/Ecuador_brochure.pdf

Boas, H. (Ed.) (2011). *The No REDD papers volume I*. Portland: Eberhardt Press. Retrieved from http://gjep.wpengine.com/wp-content/uploads/2011/11/noreddpapers.pdf

Böhm, S., Murtola, A. M., & Spoelstra, S. (2012). The atmosphere as business. *Ephemera: Theory & Politics in Organization*, *12*(1/2), 1–11.

Bond, P. (2012). Market failure at Durban's climate summit. *South African Geographical Journal*, *94*(2), 89102.

Bremer, L. L., Farley, K. A., & Lopez-Carr, D. (2014). What factors influence participation in payment for ecosystem services programs? An evaluation of Ecuador's SocioPáramo program. *Land Use Policy*, *36*, 122–133.

Büscher, B., & Fletcher, R. (2015). Accumulation by conservation. *New Political Economy*, *20*(2), 273–298.

Cabello, J., & Gilbertson, T. (Eds.). (2012). A colonial mechanism to enclose lands: A critical review of two REDD+-focused special issues. *Ephemera: Theory & Politics in Organization*, *12*(1/2): 162–180.

Cavelier, J., & Grey, I. M. (2014). *Payment for ecosystem services schemes*. Retrieved from http://thegef.org/sites/default/files/publications/28252nomarks_0.pdf

CfRN (n.d.). *Coalition for Rainforest Nations: Coalition nations*. Retrieved from https://www.rainforestcoalition.org/coalition-nations/

Climate Alliance (n.d.). *About us*. Retrieved from http://www.climatealliance.org/about-us.html

COICA. (2008). *Informative manual on climate change for the indigenous peoples of the Amazon Basin: REDD and its implications*. Quito: COICA.

COICA. (2013). *Alternativa de REDD+ Indígena Territorios de Vida Plena para Enfriar el Planeta*. COICA-WWF: Perú - Ecuador.

COICA. (2015). *Presentation of the coordinator of indigenous organizations of the Amazon Basin to the subsidiary body for scientific and technical advice (SBSTA), at its 41st session*, Unofficial Translation. Retrieved from https://unfccc.int/sites/default/files/469.pdf

Corbera, E., & Kosoy, N. (2010). Payments for ecosystem services as commodity fetishism. *Ecological Economics*, *69*(6), 1228–1236.

Corbera, E., & Schroeder, H. (2011). Governing and implementing REDD+. *Environmental Science & Policy*, *14*(2), 89–99.

Corson, C., & MacDonald, K. I. (2013). Enclosing the global commons: The convention on biological diversity and green grabbing. *Journal of Peasant Studies*, *39*(2), 263–284.

Dauvergne, P. (2016). Environmentalism of the Rich. Cambridge: MIT Press.

Dehm, J. (2012). REDD faces all around': Implementing reducing emissions from deforestation and forest degradation in Indonesia. *Local-Global Journal*, *10*, 98–125.

Fehse, J. (2012). Private conservation agreements support climate action: Ecuador & Socio Bosque Programme. Climate and development knowledge network (CDKN). *Inside stories on climate compatible development, 6*. Retrieved from https://www.gov.uk/dfid-research-outputs/private-conservation-agreements-support-climate-action-ecuador-s-socio-bosque-programme

Ford, L. (2012, June 21). Rio+20 politicians deliver 'new definition of hypocrisy' claim NGOs. *The Guardian*. Retrieved from https://www.theguardian.com/global-development/2012/jun/21/rio20-politicians-hypocrisy-ngos

Furtado, F. (2018). REDD+, The Carbon Market and California-Acre-Chiapas Cooperation: Legalizing Mechanisms of Dispossession. Retrieved from https://www.foei.org/wp-content/uploads/2018/01/REDD_Thecarbon-market-and-the-California-Acre-Chiapas-cooperation.pdf

Gerlach, A. (2003). *Indians, oil, and politics: A recent history of Ecuador*. London: Scholarly Resources.

Gilbertson, (2017). Building Solidarity Against the Threat of Linking Global Carbon Markets (Carbon Pricing: A Critical Perspective for Community Resistance - 1). Retrieved from: http://www.ienearth.org/wpcontent/uploads/2017/11/Carbon-Pricing-Final-Print-Final-HiRez.pdf

Hames, R. (2007). The ecologically noble savage debate. *Annual Review of Anthropology*, *36*, 177–190.

Harvey, D. (2011). *The enigma of capital: And the crises of capitalism*. New York: Profile.

Harvey, F. (2018, May 22). Carbon markets back from the brink of collapse, says World Bank. *The Guardian*. Retrieved from https://www.theguardian.com/environment/2018/may/22/carbon-markets-back-from-the-brink-of-collapse-says-world-bank

Hübenthal, A., Starnfeld, F., & Carrión, D. (2010). *The forests dialogue: Field dialogue on REDD readiness in Ecuador 15–18 June 2010*. Background Paper. Papallacta: Ecuador. Retrieved from https://theforestsdialogue.org/sites/default/files/tfd_redd_ecuador_backgroundpaper_eng.pdf

IISD (2015). Rio Conventions Pavilion (RCP) 1 December - 11 December 2015, in Paris and France. Retrieved from http://enb.iisd.org/climate/cop21/cbd-rcp/7dec.html

IISD (2016, December 18). Summary. Rio Conventions Pavilion Bulletin, 200(39), pp. 1–30. Retrieved from http://enb.iisd.org/download/pdf/sd/enbplus200num39e.pdf

ILRC. (2012, October 26). An open letter to the co-chairs of the UN-REDD programme policy board [open letter]. *Indian Law Resource Center.* Retrieved from http://indianlaw.org/sites/default/files/ILRC%20letter%20UNREDD%20Policy%20Board%20re%20Ecuador%20NP%20Oct%20%202012.pdf

IPCCA. (2011). Report on the local evaluation of the binational territory of the Zapara nation in the Ecuadorian Amazon. Asociación de Mujeres Sápara (Association of Sápara women) in collaboration with land is life. Cusco: Peru.

Klagge, B., & Zademach, H. M. (2018). International capital flows, stock markets, and uneven development: the case of Sub-Saharan Africa and the Sustainable Stock Exchanges Initiative (SSEI). *Zeitschrift für Wirtschaftsgeographie*, 62(2), pp. 92–107.

Lang, C. (2009, August 11). *Indigenous people in Ecuador reject REDD.* REDD Monitor. Retrieved from http://www.redd-monitor.org/2009/08/11/indigenous-peoples-in-ecuador-reject-redd/

Lang, C. (2012, November 6). *Ecuador's conflict between oil extraction, indigenous rights and REDD.* REDD Monitor. Retrieved from https://redd-monitor.org/2012/11/06/ecuadors-conflict-between-oil-extraction-indigenous-rights-and-redd/#en

Lawson, J., Maginnis, S., & Suarez, L. (2010). Proceso de Preparación Para REDD-plus en Ecuador. *The Forests Dialogue: Resumen de la Iniciativa*, *3*. Retrieved from https://portals.iucn.org/library/sites/library/files/documents/2010-104.pdf

Liverman, D. (2004). Who governs, at what scale and at what price? Geography, environmental governance, and the commodification of nature. *Annals of the Association of American Geographers*, *94*(4), 734–738.

MAE. (2012a). *The socio bosque program and free, prior and informed consultation and consent.* REDD+ Social & Environmental Standards, Quito: Ministerio del Ambiente.

MAE. (2012b). Plan de Inversión. *Socio Bosque Programa de Protección de los Quito: Ministerio del Ambiente.*

Mato, D. (2000). Transnational networking and the social production of representations of identities by indigenous peoples' organizations of Latin America. *International Sociology*, *15*(2), 343–360.

Miles, T. (2017, May 25). U.N. calls for global economic makeover to replace neo-liberalism. *Reuters.* Retrieved from https://www.reuters.com/article/us-un-economy/u-n-calls-for-global-economic-makeover-to-replace-neo-liberalism-idUSKCN1BP2IR

MINAM. (2014). *Indigenous peoples present in the climate challenges. Communication1/05.10.14*. Retrieved from https://www.scribd.com/document/243873577/Indigenous-Peoples-present-in-the-global-climate-challenges

Nuñez, M. (2011). *Rights and REDD in Amazonian countries and Paraguay: Systematization of regional workshop*. Retrieved from https://d5i6is0eze552.cloudfront.net/documents/Publikasjoner/Andre-rapporter/Regnskograpport-EquadorENGspreads-1.pdf?mtime=20150630110735

O'Neill, K. (2009). *The environment and international relations*. New York: Cambridge University Press.

Pieck, S. K. (2013). Transnational activist networks: Mobilization between emotion and bureaucracy. *Social Movement Studies*, *12*(2), 121–137.

Pirard, R. (2012). Market-based instruments for biodiversity and ecosystem services: A lexicon. *Environmental Science & Policy*, *19*, 59–68.

Rainforest Foundation UK, (2017). Retrieved from https://www.rainforestfoundationuk.org/media.ashx/fcpfletter.pdf

RCP (n.d). Rio Conventions Pavilion: Linking Biodiversity, Climate Change, and Sustainable Land Management. Retrieved from http://www.riopavilion.org/

Reed, P. (2011). REDD+ and the indigenous question: A case study from Ecuador. *Forests*, *2*(2), 525–549.

Saavedra, L. A. (2011). *Ecuador: Threats of the carbon market*. Retrieved from http://ccmin.aippnet.org/index.php?option=com_content&view=article&id=720:ecuador-threats-from-carbon-market&catid=110:news&Itemid=11

Sawyer, S. (2004). *Crude chronicles: Indigenous politics, multinational oil, and neoliberalism in Ecuador*. London & Durham: Duke University Press.

Schrijver, N. (2016). Managing the global commons: Common good or common sink? *Third World Quarterly*, *37*(7), 1252–1267.

SDSN. (2015). Información Sobre la Región Amazónica en Ecuador. Retrieved from https://info.undp.org/docs/pdc/Documents/ECU/SDSN-ECUADOR_SEPTIEMBRE_22.09.2015.docx

Spash, C. L. (2015). Bulldozing biodiversity: The economics of offsets and trading-in nature. *Biological Conservation*, *192*, 541–551.

Stavins, R. N., & Stowe, R. C. (Eds.). (2017). *Market mechanisms and the Paris Agreement*. Cambridge: Harvard Project on Climate Agreements. Retrieved from https://www.belfercenter.org/sites/default/files/files/publication/2017-10_market-mechanisms-paris_v5.pdf

Terra Livre. (2012a, June 22). Cúpula dos povos por justiça social e ambiental: Contra a mercantilização da vida, em defesa dos bens comuns. Retrieved from http://blogapib.blogspot.com/2012/06/documento-final-do-ix-acampamento-terra.html

Terra Livre. (2012b, June 23). *Indigenous peoples' Terra Livre Declaration at Rio +20*. Retrieved from https://www.forestpeoples.org/en/topics/environmental-governance/news/2012/06/indigenous-peoples-terra-livre-declaration-rio-20

Tienhaara, K. (2014). Varieties of green capitalism: Economy and environment in the wake of the global financial crisis. *Environmental Politics*, *23*(2), 187–204.

UN. (1997). *Earth summit: The UN conference on environment and development (1992)*. Retrieved from http://www.un.org/geninfo/bp/enviro.html

UN. (2015). *United Nations sustainable development knowledge platform: Coalition for Rainforest Nations*. Retrieved from https://sustainabledevelopment.un.org/index.php?page=view&type=1002&nr=32&menu=812

UN Brussels. (2011, December 5). *What is "Rio+20"?* Retrieved from http://www.unbrussels.org/what-is-rio20/

UNCSD. (2012). *About Rio+20*. Retrieved from https://sustainabledevelopment.un.org/rio20

UNDP. (2015). *Financing solutions for sustainable development: Payment for ecosystem services*. Retrieved from http://www.undp.org/content/sdfinance/en/home/solutions/payments-for-ecosystem-services.html

UNDPHDR. (2011). *Human development report: Sustainability and equity, a better future for all*. Retrieved from http://hdr.undp.org/en/

UNDRIP. (2008). *United Nations declaration on the rights of indigenous peoples*. Retrieved from https://www.un.org/esa/socdev/unpfii/documents/DRIPS_en.pdf

UNFCCC. (2010). *Report of the conference of the parties on its sixteenth session. Addendum part two: Action taken by the conference of the parties at its sixteenth session. FCCC/CP/2010/7/Add.1*. Retrieved from http://unfccc.int/resource/docs/2010/cop16/eng/07a01.pdf#page=2

UNFCCC. (2016, August 9). *Indigenous peoples central to climate action*. Retrieved from https://unfccc.int/news/indigenous-peoples-central-to-climate-action

WWF. (2014). *Amazon indigenous REDD+ Launched at UNFCCC COP20*. Retrieved from http://wwf.panda.org/wwf_news/?235956/Amazon-Indigenous-REDD-launched-at-UNFCCC-COP-20

2

TAKING UP SPACE AND MAKING UP OTHERS

The Neutral Observer

The emergence of "indigeneity" as a notion has contributed to the construction of a global socio-political context within which some people are understood to be part of a marginalised and fragmented (yet somehow homogenous) international category of people who share some selection of various characteristics that are associated with being indigenous. It is not at all clear, though, what the terms "indigenous" and "indigeneity" mean. They are complex and contested, and neither is, nor ever has been, universally agreed upon (Kuper et al., 2003, p. 389; Radcliffe, 2015, p. 2). That said, assigning this identity to selves or others appears, in certain circumstances, to be done quite uncritically. Though often treated as synonymous with similar terms (including "tribes", "natives", "Indians", "locals", "Aborigines", and "First Peoples"[1]), the growing political importance of being indigenous *specifically*, particularly in relation to discussions around climate change,[2] has meant that it has been necessary for more precise cultural, political, and legal definitions to begin to created.[3] This has happened despite various Indigenous representatives – including the Special Rapporteur to the UN on Indigenous Peoples, Erica-Irene Daes, in 1995 – arguing that a definition of the concept of "Indigenous people" is neither necessary nor desirable because "historically, indigenous peoples have suffered, from definitions imposed by others" (Chen, 2014, pp. 3–4).

The very idea of *being Indigenous* has resulted in the creation of an essentialist category that is located in opposition to a generalised hegemonic mode of civilisation: being "modern", "cosmopolitan", "Western",

DOI: 10.4324/9781003341864-4

or "developed", all of which are terms that are implicitly associated with characteristics that include being "rational", "scientific", "objective", and "educated" or "professional" (Helland & Lindgren, 2016, pp. 432–438). But this opposing category, at least in mainstream political and environmental discourse, is rarely referred to as being "non-indigenous", and the two categories are not generally considered in their historical context as coloniser (non-indigenous) and colonised (indigenous). Instead, they are implicitly treated as one being "Neutral" and the other being the "Other". But being non-indigenous, which is often understood as being "White", is clearly not neutral. It simply hides itself, as Tascón (2004, p. 242) put it, within the "myriad of different ways in which people receive privilege". But despite being hidden, it still has practical effects on relationships that lead to members of the supposedly neutral category gaining further privileges, while excluding and subjugating Others (Tascón, 2004, p. 242).

The historic and contemporary role and impact of academic research in establishing these categories – Neutral and Other – are also rarely considered. The methodological approach I discuss in this chapter begins by acknowledging that the very concept of being indigenous (and therefore being *non*-indigenous) has not appeared in a vacuum. Instead, it has come into existence alongside, and has emanated from, a history of Eurocentric politics, economics, and epistemology; the very idea of indigeneity flows from within the political and academic institutions of Europe,[4] which are embedded within the remnant structures of the colonial project. As such, this book, and any research that discusses these themes, even when doing so from a critical perspective, should be situated, understood, and critiqued, within the context of this history of domination (Quijano, 2000, p. 533). Both development studies and anthropology have contributed to and sustained imperialist discourses about the Other, and this objectification through anthropological representation is connected to political hierarchies that are embedded within North-South power relations (Gardner & Lewis, 1996, p. 23). As such, this book is not neutral: it is a product of the privilege and power of *seeming* neutral, of entering and leaving the worlds of Others at will, and of escaping representation by the Other, which allows me to feel I have seen *them* without being seen (Haraway, 1988, p. 581), and to write *them* without being written.

Gal (1998) noted that various internal and external factors, including demands for recognition by Indigenous Peoples and the creation of social theory in postcolonial settings that view Europe from the outside, have resulted in anthropologists avoiding the uncritical use of apparently neutral terms without first critiquing the historical context in which they emerged (Gal, 1998, pp. 324–325). Thus, rather than studying "modernisation" or "tradition", anthropologists are now more inclined to study the "inventions of tradition" or the "invention of the modern" against the backdrop of the history of European political domination. Even so, as noted by Rigney (2006, p. 34) (see also, Smith, 1999), the discourses that began to emerge in various disciplines that speak about Indigenous Peoples have become littered with problematic issues that include:

- Research that is framed by the researcher's priorities and interest rather than the needs of Indigenous communities;
- The reduction of Indigenous ownership of Indigenous knowledges and intellectual property;
- The lack of ongoing consultation, negotiation, and involvement of Indigenous communities in the design, facilitation, and publication of research;
- Inappropriate research methodologies and ethical research processes; and
- The need for effective, appropriate, and culturally sensitive research in relation to ethics and protocols.

(Rigney, 2006, p. 35)

My research made all of these mistakes and more, and in writing it up in a way that is self-critical, giving examples of how and why I think these mistakes were made, I intend to go some way towards creating something "useful" out of what could otherwise be seen as problematic or as a "failed" research project.

Though this book is intended to be highly reflexive and critical in its approach, drawing upon theory and literature relating to feminist scholarship and critical indigenous and decolonial studies, it would be quite a stretch to call it either feminist or decolonial research, as such. Rather it can be understood as an attempt to navigate and negotiate, understand, and illustrate the limitations of my research (and of academic research more generally) through an exploration that takes these critical perspectives seriously. In the sense that it considers the social, historical, and ideological forces that contribute to constructing categories of people, my research is somewhat influenced by critical theory. But the progressive reading of history associated with critical theory represents a problem – the meta-narrative of Westernisation *as* "development" – not least in its potential for subalternising and invisiblising Other epistemes (Walsh, 2007, p. 224). Allen's (2017) efforts to overcome the neo-colonialist tendency within critical theory (which assumes that the present is the product of a history of cumulative learning and therefore represents "progress", "development", or "social evolution") is useful here. It provides a strong argument, heavily influenced by Adorno, but drawing on decolonial scholars including Chakrabarty, Mignolo, and Quijano, for a decolonised critical theory as way of achieving emancipatory social goals:

> If we are to hold on to the idea of progress as a forward-looking moral political imperative, that commitment will have to go together with a relentless and ongoing problematization not only of any and all judgments about what would constitute progress but also of the normative standards by which such progress could be measured.
>
> *(Allen, 2017, p. 228)*

Allen suggests that an emancipatory imperative is a characteristic shared by critical theory and decoloniality. But she argues that for critical theory to correspond with decolonisation, it is necessary to shift the geography of knowledge, to learn to unlearn, and to seek a genuinely open, and open-ended, dialogue with colonised or subalternised subjects. In turn, this requires an epistemic humility regarding one's own normative commitments (Allen, 2017, pp. 209–210). I am not sure how possible it is to *really* be "epistemologically humble", but this book, at least at the personal level, is intended to represent a process (however limited) of unlearning[5] and to demonstrate an openness to dialogue.

The Spectre of Re-colonialism

Notions of "progress", what it means and how it can be achieved, underpin the idea of "development", but this provides little in the way of a meaningful definition of the term. Over 30 years ago, in *The Development Dictionary: A Guide to Knowledge and Power*, Sachs described "development" as a towering lighthouse, guiding the nations that were newly freed from the subordination of colonialism on their journey through post-war history. But he noted that by the final decade of the twentieth century, cracks had begun to show, fissures of delusion and disappointment, until the lighthouse began to crumble into the sea. "Development", he acknowledged, had not worked, and it now "stands like a ruin on the intellectual landscape", exhausted, outdated, and obsolete (Sachs, 1993, p. 1). And yet, throughout my undergraduate degree in Latin American Development Studies (which began almost 15 years after Sachs's observations were made), and my master's degree in International Relations, this crisis in development rarely cropped up. There seemed to be an implicit assumption that development *was* problematic but that this had been recognised and it was now, supposedly, being done better. Very little time was spent considering the historic role of colonialism, and even less attention was paid to the continued role of colonial power in development and how this shapes what knowledge, and whose knowledge, "matters".

Though I was attracted to studying development as a means through which to understand global inequality and injustice, I have always been suspicious of development as a practice and as an economic model. But I was somewhat convinced of the value of development *studies* as a critical process through which to better understand how and why some people win and others lose as development is unrelentingly pushed forward in an apparent attempt to better the world. However, during my doctoral studies, two significant changes occurred in my understanding of development studies. The first of these, as noted above, is regarding the role of international development (both the industry and the field of research), as well as anthropology, in creating and recreating categories of people as subjects of study, assigning them characteristics (usually based on academic understandings of race, class, nationality, gender, and so on), and defining their value (usually based upon how useful they are for research or in achieving development "goals"). The second was my naïve assumption that development and the study of it was (however slowly) getting "better". This second idea was challenged by a 2017 article, now retracted by the publisher (Taylor and Francis), in which a professor of International Development - a supposed development "expert" who happened to be a white Western man - made his "case for colonialism".

I am not going to reference the article directly because, despite being retracted, the controversy surrounding it has meant that it has already been excessively cited.[6] In the current academic culture, this affords it a degree of legitimacy and recognition[7] that it does not, in my opinion, deserve. As this book is largely about which ways of knowing and being are legitimised, there is no need to bolster the article's notoriety further. It was not particular interesting, but what was interesting was the reaction to it. Responses ranged from tepid and careful endorsements (which usually argued that colonialism was bad, but that the world without it would have been worse), through eye rolling and shaking of heads, to outrage and anger.[8] But the very fact that it was published at all, especially in a journal whose first issue in 1979 stated that "Our concern is the Third World: we will speak for it, indeed, speak with its own voice", served as a reminder that, for some white men, development and colonialism are not just intertwined but rather that they *should* be, and that for this particular journal, this view counts as speaking *for* the "Third World". Moreover, once Taylor and Francis decided to retract the article, it was not because the editor of *Third World Quarterly* had a sudden change of heart or realised that a problematic and sub-standard piece had slipped through the editorial net.

Instead, on the webpage where the article could once be found, there is now a statement that reads:

> Taylor & Francis conducted a thorough investigation into the peer review process on this article. Whilst this clearly demonstrated the essay had undergone double-blind peer review, in line with the journal's editorial policy, *the journal editor has subsequently received serious and credible threats of personal violence* [emphasis added]. These threats are linked to the publication of this essay. As the publisher, we must take this seriously. Taylor & Francis has a strong and supportive duty of care to all our academic editorial teams, and this is why we are withdrawing this essay.
>
> *(Taylor & Francis, 2018)*

That the essay making a "case for colonialism" was *not* seen as part of an ongoing process of marginalisation, subjugation, and violence, and that the publisher considered its "duty of care" to be solely to its academic team points to a wilful ignorance of the writer and the editor, but it also raises concerns regarding the established field of development studies.

(Not) Doing Decolonial Anthropology of Development

With the above concerns in mind, I position my research within the field of anthropology of development and alongside decolonial studies, but I acknowledge that my fieldwork was largely unsuccessful in practicing decolonial methods and that the book itself remains embedded within (and is therefore limited by) the epistemological and ontological framework of Western academe. As Sillitoe noted:

> The extent to which it is possible to escape our intellectual frame of reference has become a central topic of debate, particularly with the advent of postmodernism, which argues that anthropology's ideological (or theoretical) concerns inevitably distort its subject matter. It is likely that jointly exploring and reviewing indigenous critiques will help address this problem; the indigenous scholar is an integral party, making us aware of issues and helping us understand other views.
>
> *(Sillitoe, 2016, p. 16)*

Though I agree with this statement to some degree, it is not what I did. My research, as it had to be, was designed by me in the United Kingdom in a way that would appeal to the ideological, ontological, and epistemological preferences of the institutions to which I applied for study and funding (specifically, the ESRC and the University of Sussex), not the people with whom I would conduct my research. This is indicative of a broader theoretical problem associated with conducting research about or with Indigenous Peoples, or any subalternised or marginalised peoples. The agenda, the research questions, the methodology, and the space, place, and time horizons of the research are largely dictated (either implicitly or explicitly) by the research institutions, by the preferences of departments or supervisors, and/or by the researcher. Many suggestions have been made regarding how this problem might be overcome, such as including Indigenous scholars in the design of research and in conducting it (as noted above). Smith, for example, through her own focus on Maori studies noted some of the strategies that scholars have employed in order to find a way "around the problem" and be more culturally sensitive. These include:

1. the strategy of avoidance, whereby the researcher avoids dealing with the issues or with Maori;
2. the strategy of "personal development", whereby the researchers prepare themselves by learning Maori language, attending *hui* and becoming more knowledgeable about Maori concerns;
3. the strategy of consultation with Maori, where efforts are made to seek support and consent;
4. the strategy of "making space" where research organisations have recognised and attempted to bring more Maori researchers and "voices" into their own organisation.

Along with "the strategy of partnership, whereby the organisation recognises the need to reflect partnership at governance level and embed it in all its policies and practices", these approaches all have potentially positive and negative consequences for the researchers and the researched (Smith, 1999, p. 179). My concern relating to indigenous studies in general, but particularly regarding my own research, is that even if any or all of these approaches are adopted this is only done in situations where the principal (usually non-indigenous) investigator holds the opinion that this *should* be done, and in many instances *the research collaborators will be chosen based upon how well they fit with or within this ideological perspective.*

What this means is that those marginalised people who have been best assimilated into academia or who best represent a preconceived notion of who and what *should* be considered important in academia are prioritised when selecting which Others are most useful. This in turn legitimises the research that is conducted by virtue of demonstrating "participation" and simultaneously legitimises the views and identities of *some* Indigenous people. I am aware that this argument folds in on itself, and I am glad that it does. I consider this to be the hard problem of writing about people who might be described as marginalised or subaltern: how can research be conducted on or with "them", their knowledge, and their perspectives, in a manner that does not perpetuate processes of domination, colonialism, and subjugation? My first thought is that the question also has a hard answer: it cannot. It just might be the case that academic research, due to the historic context within which it emerged and its contemporary position in the hierarchy of knowledges, is fundamentally and inescapably limited in what it can tell us about the world, particularly the world of Others. This is not a new concern. In *The Scope of Anthropology*, Lévi-Strauss acknowledged the position of anthropology (and anthropologists) in a history of colonialism:

> If society is in anthropology, anthropology is itself in society: it has been able to enlarge progressively its study to the point that of including therein the totality of human societies: although it has appeared at a late period in their history and in a small sector of the inhabited world. More than that, the circumstances of its appearance are comprehensible only in the context of a particular social and economic development: one suspects, then, that they go together with a dawning awareness – almost remorseful – that humanity could have remained alienated from itself for such a long time, and above all, *that that fraction of humanity which produced anthropology should be the same fraction of humanity to have made so many other men the objects of execration and contempt. "Sequels to colonialism", it is sometimes said of our investigations* [emphasis added]. The two are most certainly linked, but nothing would be more misleading than to see anthropology as a throwback to the colonial frame of mind, a shameful ideology which would offer colonialism a chance of survival.
>
> *(Lévi-Strauss, 1966, p. 51)*

Nonetheless, I do not consider this to mean that the pursuit is entirely in vain, only that if there is a hard limit to what can be achieved through academic study, then emancipatory social goals (if that is indeed the aim of a *forward-looking* notion of "progress") might be better addressed elsewhere. I do, however, also happen to believe that the limit has not yet been reached: there is still plenty of work to be done. But in order to make new contributions to knowledge within the framework of Western Academia, particularly in relation to research that focusses on power relationships, it is necessary to go beyond simply paying lip service to notions of self-critique and reflexivity, whereby one acknowledges then promptly forgets about positionality and power, and instead engaging in a "meaningful" and ongoing reflexive process.

Reflecting on Reflexivity

The general observations noted above inform and underpin much of this book, including the methodological approach, which in the first instance is intended to be a highly reflexive and self-critical process. Wasserfall described the distinction between "weak" and " strong" reflexivity as follows and highlights a risk in taking this approach:

> the "weak" reading of reflexivity is a continued self-awareness about the ongoing relationship between a researcher and informants, which is certainly epistemologically useful: the researcher becomes more aware of constructing knowledge and of the influences of her beliefs, backgrounds and feelings in the process of researching. Reflexivity is a position of a certain kind of praxis where there is a continuous checking on the accomplishment of understanding [...] The "strong" reading, on the other hand, contains certain assumptions or the deconstruction of the authority of the author and/or of the power difference in the field. These assumptions gloss over difficult theoretical and political tensions in which ethnographic knowledge is produced and consumed.
>
> *(Wasserfall, 1993, pp. 24–25)*

A critique of strong reflexivity, then, is that it could be interpreted as the author seeking to escape responsibility or authority, and it could even be used to suggest that the objectivity of the research has been enhanced (see, e.g., Pels, 2000, p. 12) and so it is important to not consider reflexivity (in either the weak or strong reading) to be an easy fix for the limitations of a partial perspective. The goal of reflexivity is not to present findings, and it does not provide final "results". Instead, it evokes a response within the reader and offers observations that remain open to interpretation (Ploder & Stadlbauer, 2016, p. 754). When it works well, a reflexive approach that makes writing more personal can be considered a course correction that contributes to moving anthropological research away from, and supposedly beyond, the imaginary objectivist/realist observations of the past (Riessman, 2015, p. 228). It is for this reason that reflexivity must be accompanied by the "unlearning" of one's own privilege and doing what Spivak refers to as "homework" (1990, pp. 62–63). Thus, it is neither my intent to liberate myself from the burden of partiality or subjectivity nor to claim that my perspective is more accurate because I have taken this approach. Rather,

I am trying to acknowledge and expose (to the extent that I can *from* my position) the sites of power within which the researcher, the researched, and subsequently the research I have produced, are located. As Presser (2005, p. 2087) noted:

> The researcher's goal is not to emancipate *the* authentic story of the narrator – none exists – but rather to expose as much as she can of the relations that influence the construction of the story that is told.

My research was not designed in this way: it could not have been. My understanding of the role of my research, and of myself within it, changed considerably throughout the course of my fieldwork and in writing-up, as did the types of knowledge I value. This ultimately led me to question the legitimacy and validity of conducting research with Indigenous Peoples in the manner I had intended, or, indeed, doing so at all.[9] For example, when I wrote my research proposal, I did not expect to be writing about the relationship between research and colonialism, at least not to the extent that I have, though I did acknowledge that the relationship existed. Rather, my experiences during and after fieldwork magnified my peripheral concerns relating to researcher positionality, power, and identity. This aligned with my existing feelings of reticence and resentment towards academia, which is rooted in my experience of being a working-class student, researcher, and teacher in a system that I am highly sceptical of. As such, an important part of my methodology is to retrospectively critique my ethnographic research, which (though it is far from unique in this respect) I consider to have been highly problematic, and to consider why *I* feel so conflicted when it comes to research of this kind.

Research Aims and Questions

Taking a self-critical and reflexive approach also means that my research questions have been revisited and have changed quite considerably. As such, the book I have produced can be understood as a snapshot of where my understanding happened to be at the moment of writing. This means that any conclusions drawn are limited not only by the partial nature of the data collected but also by the temporal position within which it was written up. Thus, though the themes have remained largely the same, the overarching questions that now run as threads through this book are as follows:

1. What does it mean to be indigenous (and non-indigenous) in spaces where climate change is being discussed?

2. What, if anything, do Indigenous people "know" about climate change, and what role does this have in shaping responses to it?
3. What impact does research (including my own) have on the researcher and the researched?

Each of these questions is, in some way, related to the broader theme that this book considers, which is how power and knowledges flow through the social body (see Foucault, cited in Fraser, 1981, p. 272), and what this means for the construction of identities or categories of people. The aim of this book is *to provide a retrospective ethnographic and autoethnographic account of multiple fieldsites positioned at multiple political levels (ranging from the international to the personal) where I encountered and interacted with notions of indigeneity against the thematic backdrop of climate change mitigation.* Later, I will discuss how I approached gathering and analysing "data' in seeking answers to the above questions, but first, I wish to draw the reader's attention to two issues that created problems for my research: "me" and "research".

So, Enough About You, Let's Talk About Me

In his autobiographical account of becoming a historian of colonial Britain in the United States, Greene (2004, p. 18) cited a derisive "joke" attributed to Stewart Schwartz, also a colonial historian (but of Latin America) from the United States: "What did the postmodern anthropologist say to his informant? But enough about you, let's talk about me". Greene considers it "uncongenial" that, in this postmodern world "scholars find themselves compelled to spell out their personal relationship to their subjects of study", primarily because he considers scholars to be uninteresting people who live mundane lives, spending their time teaching or sitting around in libraries, reading, and thinking about "intellectual problems". To my mind Greene's misguided assertion illustrates a problem that has been pervasive in the history of academia, which is precisely what has required the generation that has followed to take pains to untangle the mess of privilege and power that has been left in his generation's wake.

To consider the researcher as irrelevant and separate to one's research imagines the self as an objective, rational observer, who simply ponders such intellectual problems from outside. "We" are uninteresting; "they" are interesting. The scholar is neutral, and the subject is the problem, but being a problem, as W. E. B. Du Bois noted on the first page of *The Souls of Black Folk*, "is a strange experience, – peculiar even for one who has never been anything else" (Du Bois, 1903, p. 1). As an African-American man and a prolific sociologist, historian, and Pan-Africanist civil rights activist in the 19th and 20th centuries, Du Bois both knew and felt what it was to be a "problem". And this idea, that the lives of some people are neutral and irrelevant while the lives of Others constitute a problem to be solved, is indicative of the underlying politics of social research, including development anthropology, which Escobar (1991, p. 677) described as a "politics of a Western-based, patriarchal, scientific, economic, and cultural project". It is necessary, then, to be explicit about my position *in* my research as well as my own influence *on* my research, acknowledging that my identity is not outside of this history or outside of this book, but that they are intrinsically linked to the extent that they are almost indistinguishable from one another. A tool of reflexivity that is useful for untangling this relationship is autoethnography, described by Ploder and Stadlbauer (2016, p. 754) as a process whereby the researcher is required to "actively engage with her story, reflect her affects, physical experiences, mental states, anxieties, joy, and excitement, and work on their textual or performative representation. In short, she needs to work with and through her story and text". But it is not enough to simply tell one's own story, rather moments of knowledge production are triggered when the material is effectively "worked through" from a critical perspective (Ploder & Stadlbauer, 2016, p. 754). My relationship to my research is worked through throughout the substantive chapters of this book (particular in the final chapter), but it is also pertinent here, as it has shaped even my methodological approach.

*

I was a "problem", once: I was a white working-class boy in Britain. McDowell (2011, p. 7) described this category as "a complex and hybrid group whose ethnicity and gender – as white men – endows privilege in certain spheres, but whose age and class position – as working-class adolescents – locates them as subordinate". As Theresa May noted in her first speech as Prime Minister (with what I found to be unconvincing concern), being a white working-class boy situated me in the group least likely to go to university in Britain (May, 2016). I have always felt like an outsider in higher education and have perceived the academy as an oppressive and frustrating space. The middle-classness of universities, especially when people are talking about poverty as an abstract concept that happens somewhere else, and when the term "working-class" is used as a euphemism to describe the poor, the disenfranchised, and marginalised of society – those who need to be dragged up to the respectable level of the middle-class – fuelled my cynicism of the academic system. And yet, I understood its power and wanted a part of it. Though it began as a *feeling*, the present articulation of this critique of the very position that I am now in is influenced by Bourdieu's *Homo Academicus*, in which he aimed to trap "the supreme classifier of classifiers in the net of his own classifications":

rather than domesticating the exotic, he aimed to exoticise the domestic (Bourdieu, 1984, pp. xi–xii).

Two years after I graduated, I found myself walking home from a social event with one of my undergraduate tutors. We were talking about teaching first-year students and I asked what her first impression of me had been, she laughed and said, "You were an arrogant little bastard!". I wasn't surprised to hear this; I knew I had been a difficult student at times. But I explained to her that rather than being arrogant, I had been terribly anxious and self-conscious about my own abilities. I also harboured (and still do) resentment towards higher education, which I had come to see as a necessary tool of my own social mobility, but still an elitist part of an unjust and inequitable society. Though by this time I had recognised that my self-perceived academic inadequacies were the result of structural inequalities, rather than my own "failings", I was also highly sensitive to the knowledge that overcoming this required more academic, economic, and emotional work than would be expected from middle-class students, not to mention navigating the reconstruction of my dual identity as a "working-class academic".

This highlights a potential obstacle in discussing the overlapping and contradictory advantages and disadvantages of my experience while conducting the research for this book. In academia, the apparently "normal" characteristics of my identity are being white and being male, while the "different" part is being working-class. Conversely, while working with Indigenous Peoples, being white and British (and at times being male) made me very much an outsider, but the embodied experience of having grown up in a context of relative poverty brought with it some awareness of, and sensitivity to, marginalisation.

I noticed this dual perspective during my fieldwork: few other (Indigenous or non-indigenous) people seemed to feel as uncomfortable as I did with the way that indigeneity was essentialised, displayed, and commodified. Or perhaps they just hid it well. For example, when I first visited the Indigenous Peoples Pavilion at COP20 in Lima, I left feeling unsettled, indignant, and a bit angry about what I had observed. In my fieldnotes I wrote:

> It felt like it was a zoo, where people were walking around admiring the weird and wonderful exotic indígenas, taking photos of them and asking them to pose in selfies […] I feel so uncomfortable going up to people and talking to them. I don't know what to say. Everyone else seems to do it so easily, even those who speak Spanish far worse than I do

The comparison to a zoo might indeed seem cynical, but when the history of Indigenous Peoples' presence in non-indigenous spaces, particularly exhibition spaces, is traced back far enough, it is possible to see the continuity of some undesirable characteristics. The path of this history began with the early Victorian displays of colonised people in "freak shows" (see e.g., Davies, 2016) and "Ethnological shows" or "human zoos" (see e.g., Blanchard, 2008) throughout the nineteenth century where "savages" were displayed as wonders, oddities, or monsters (Trupp, 2011, pp. 139–149). This then became sanitised in the early incarnations of the International World's Fairs in Europe and the United States where, in the late nineteenth and early twentieth centuries, people were brought from the colonies and exhibited not as animals, savages, or biological oddities, but instead as colonial subjects in their "natural" habitats, including representations of their cultural spaces and architecture (Morton, 2003, p. 1).

Later, this developed into a "celebration" of the cultural uniqueness of the colonies and of the colonised people, most notably in the Paris Colonial Exposition in 1931. There, the architects of the space attempted to create a hybrid of the French Metropole and the colonial outposts, and to portray a specific message: Maréchal Hubert Lyautey, the lead organiser, made his "case for colonialism" by describing the event's didactic demonstration of the contemporary world order that demonstrated how "colonial action, so long misunderstood, deformed, sometimes shackled, is a constructive and beneficial action". In its construction, according to architectural critic Marcel Zahar, it would "mark the memory of explorations and conquests with rare documents and masterpieces, to present exotic treasures, to prove by diagrams and photos the progression of the beneficial French effort." But it also, as Morton noted, served a secondary purpose: it was a transgressive site where normative social roles could be abandoned in favour of ordinarily taboo activities (Morton, 2003, p. 2). This was not the explicit intention in Paris in 1931 nor was it in Lima in 2014 or in Paris in 2015, but the taboo act of observing tribal people in a simulacrum of their "natural habitat" remained, and it made for an uncomfortable encounter, though everyone else seemed fine with it. I found the display of Otherness at the conferences quite distasteful, and it took me some time to understand why it bothered me so much.

Throughout my life, I have seen various cultural depictions of working-class life in Britain. On the one hand, working-class people are viewed as lazy, unintelligent, and dangerous, while on the other, we are held up as paragons of what it is to be British: hardworking underdogs - up against the odds, but stoic in our manner - with a rich and vibrant culture from which the more privileged classes can cherry pick for entertainment purposes.[10] As such, my position as a white male researcher in such places makes me complicit in (re)creating this voyeuristic, essentialising and Othering encounter, while my hypersensitivity to inequality and injustice, which is rooted in my class identity, affords me some insight into how problematic the whole thing is. Knowing this, however, doesn't make navigating social interactions in the space any easier.

Outsider-Outsider Research?

Making my identity explicit, then, is a necessary step towards unmasking the interactions between various personal characteristics that have shaped my experiences (Case, 2012, pp. 79–80; David & Wildman, 1994, p. 886), which helps me better understand my position in relation to my research. Entering the industry from a less privileged class position has meant that I have always considered my studies and my research to perform two contradictory roles: they are tools of both social mobility and of social control. Though it might seem antagonistic, I consider it important to state this position in relation to both my study and my own research because I have noticed how uncomfortable it makes other academics. In much the same way that white people seem to recoil instinctively when we are told that we benefit from the inherent structural racism in society, so too do (some) middle-class academics when confronted with the elitism of *their* "egalitarian" academy (Archer, Hutchings, & Ross, 2005, p. 8). This is especially the case when a white, British man identifies himself as an outsider. My identity as a researcher, then, is complex: it is both privileged and not. And as such my experience of conducting research has been contaminated by anxieties relating to an internal conflict between attempting to ensure my own social mobility (i.e., by getting a PhD) while also trying to avoid contributing to social oppression of Others (through speaking *for* or *about* them). Although all of this is in the hope of doing some good, or at least "doing no harm", none of it should be read as a false epistemic humility or an attempt to cast a moral safety net that absolves me of my accountability for the implications of my research. What my undergraduate lecturer saw as arrogance is related to my tendency to hold highly critical and strong opinions about topics that matter to me, and I take full responsibility for the decisions I have made in the process of conducting my research and writing-up, including those that I now consider to be "mistakes" or "failures".

I have chosen to write myself into my work through a process of selective autoethnography, making myself, my experiences, and my characteristics as a researcher explicit. In instances where I feel I was an "insider", I write myself in, and at other times, when I feel I was more outside of the context, looking in, I write myself out. In doing so I intend to illustrate that my presence in some places and spaces carries more weight and disrupts the fieldsite (at least from my perspective) to a greater degree than in others. This includes noting my personal characteristics, in particular as a "white man" where I feel this matters and using the term "non-indigenous" in contrast to "indigenous" in order to avoid the implicit neutrality of *not* being indigenous. For example, in the COP conferences discussed in my first empirical chapter, I am present in the text as one of many white men in a space that is indigenous in a larger place that is not. As such, I had relatively little impact on changing the context itself, other than in reinforcing existing characteristics of it in relation to indigenous/non-indigenous interactions.

There, I am, in some sense, an insider because it is a shared space where being non-indigenous and being a researcher constitutes a part of what makes it, potentially, a pragmatic, mutually constructed and mutually comprehensible "middle-ground" (Wright, 1991 cited in Conklin & Graham, 1995, p. 695), or a "hybrid" or "third space" (Bhabha, 1994). And so, I include myself in some of my descriptions from a third-person perspective in order to illustrate how I viewed my being there in retrospect and as a part of the process that I am critiquing. I recognise, though, as Barthes noted in 1977, that doing so could be viewed as obscurantist or even nefarious: "To speak about oneself by saying 'he' can mean: *I am speaking about myself as though I were more or less dead*...or again: I am speaking about myself in the manner of a Brechtian actor who must distance his character: show rather than incarnate him" (Barthes, 2010, p. 169). I do intend to "show" myself, here, in the same manner as I "show" Others, and I hope this encourages reflection on what isn't usually shown. This is done in a quite matter of fact way, reserving emotional involvement for the later chapters in which I recount my personal internal experiences more directly.

When in the indigenous community where I conducted workshops, my presence is more clearly alien to the space as I was the only white Western man (even when I was there with other outsiders, they were either women or were Ecuadorian, or both, which means my presence had a different impact). There, I was undoubtedly an outsider, and so I treat myself as such, writing my observations from my perspective in a position where I felt like a tolerated interloper. My positionality is particularly salient in the chapter that focusses on fear, where I discuss the emotional impact of encountering violence during fieldwork and how this influenced my understanding of the fear I observed in Others. As such, I look inward, reflecting on my thoughts and my emotions in relation to "the field" and the people with whom I interacted. Ultimately, this period of introspection led me to a fundamentally different understanding of what constitutes "knowledge", and this in turn reinforced my critique of academic research. It reminded me once again that it is, and can only ever provide, a partial perspective and that its self-appointed position at the top of an epistemological hierarchy is part of a process of intellectual domination and control.

Substantial research and discussion relating to insider and outsider research have been conducted, which has shed light on the strengths and weaknesses of both,[11] and so I do not consider it necessary to discuss these in detail here, and instead refer the reader to the work of Johnson-Bailey et al. (2001), Rabe (2003), and Lins Ribeiro (2016). Likewise, the space between these two positions – the insider/outsider – has been discussed at length in existing literature, including Buckle and Dwyer (2009) and Kerstetter (2012). What I do wish to highlight is my complex positionality in relation to the world of research in general, and to my research specifically, where I consider myself to be, in some sense, an outsider-outsider. What I mean by this is that in most of my fieldsites, by virtue of being a white Western man and/or by not being a direct participant in most of the interactions I observed, I was an outsider. But I also consider myself to be an outsider in the institutions of academia for and within which my research is conducted. I am clearly not "inside" in relation to the specific subject(s) that I am writing about (Indigenous Peoples and indigeneity), but neither has my lived experience always been embedded in the institutions and social structures of Western elitism where I now find myself, both in my fieldsites (at UN conferences, for example) and in universities. Moreover, the latent resentment and the suspicion I feel towards the world of academia means that I tend to not only *see* privilege, power, and inequality but also to *feel* indignant and often angry about it. This brings with it a perspective that could be viewed as a potential strength and/or weakness. I have a motivation and ability to see my own research from a highly, and potentially excessively, critical standpoint, but at the same time, I must take care to not consider my position to be equal to the "subjects" of my research. I recognise that it is, in the end, my voice that is documented here, and even when the voices of others are included, I have selected them carefully.

Power and the Authoritative Voice

Power differences and authority are central to my critique of how and why Indigenous Peoples are included in or represented in the process of negotiations around climate change, so it would be negligent to not consider the research I have produced in the same terms. In the subsequent chapters, I refer to climate change negotiations in terms of structural and epistemic violence against Indigenous Peoples and other marginalised groups, and to shy away from the knowledge that my own research is also a part of this process would be intellectually dishonest and highly problematic. This is in part due to the way in which my research was conducted and my position of relative power in producing it, but it is also due to structural and epistemic limitations that I am subject to. As Ruth (2008, p. 100) put it, "The power to represent society as is or as desired is differentially distributed. There is the authority to ensure that one's authorship is recognised as the official version, and there is the authority to have one's version recognised as an authentic version". This asymmetrical distribution of power, in many senses, but particularly in relation to the knowledge produced, is often heavily weighted in favour of the researcher rather than the people who the researcher (re)presents. The aims and ambitions of anthropologists are shaped, according to Whiffen (2004, p. 1), by an ongoing desire to legitimise anthropology itself, as a "useful" academic discipline that is worthy of respect and, crucially, funding. In *Anthropology and the Development Encounter* – which built upon Asad's *Anthropology and the Colonial Encounter* that noted a remnant "reluctance on the part of most professional anthropologists to consider seriously the power structure within which their discipline has taken shape" (1973, p. 5) – Escobar (1991) questioned the future of "development anthropology" arguing that:

> ... the practitioners' view of their own future calls for a further integration into the development apparatus. I would rather argue for a type of anthropological practice that distances itself from mainstream development institutions and conceptions, even when working within the "development" field; a type of practice that is sensitive to the remaking of social analysis that critical anthropologists seem to be working toward; a type of practice that is less concerned with standard anthropological problems and more concerned with, for instance, social movements, political struggles, and *the reconstitution of identities through development technologies and resistance to them* [emphasis added]; a type of practice that considers detached objectivity only one intellectual method among many and that, more generally, applies sustained epistemological pressure to conventional scientific anthropology and the development encounter practices and divisions of labor. A type of practice, finally, that is not threatened by otherness and difference, trying to conceal them in the displacements of its discourse (seeing others as "underdeveloped" or "needy"), but that, always aware of the power dynamics at play, searches for a more self-aware communication among different, yet equal, subjects.
>
> *(Escobar, 1991, pp. 677–678)*

In light of this observation, the limitations of academic research are often readily acknowledged, but this self-critique is softened by a general tendency to claim that the research being done *challenges* these limits.

Even the quote from Escobar above does this to some extent, because while it criticises development anthropology, it holds on to the idea that it can and should become a different (better) type of practice. Another example of this is Johnson (2008), himself an indigenous man, who begins his article on the "Tricky Ground of Indigenous Research" with the following statement:

> I wish to start this article in search of a middle path; there will be no effort made to hide myself behind some outdated and outmoded convention that pretends a disembodied and objective author/researcher has produced this work. I free myself from (the myth of) objectivity and follow in the path of feminist and critical researchers by recognizing and identifying my own positionality. I hope, through writing this article, to build an atmosphere of safety "where I can begin to speak from an integrated place" as an indigenous man and not just as a social scientist "who normally speaks as an objective authority and removes [himself] from the spoken or written word."
>
> *(Johnson, 2008, p. 127)*

The middle path Johnson describes is intended to negotiate a space for as much inclusion as can be accommodated within academic research and writing, and to push the limits of it (Johnson, 2008, p. 134). Though this is undoubtedly important, particularly when being done by self-identified Indigenous scholars, such attempts could also be seen as implying that academic research is potentially unlimited, if it can only be done right. In the field of subaltern studies this tension has been identified: Beverley (1999, pp. 31 and 38) acknowledges the difficulty of attempting to develop new forms of knowledge and pedagogy within academia while simultaneously attempting to critique academic knowledge "as such" and the "fundamental inadequacy" of the institutions that contain it.

In the contemporary context, it might be more useful to acknowledge that there are ways of being and knowing that cannot be understood (by "us" at least) through this process, *and that that is ok*. And it does not mean the process itself is not useful or important.[12] This might be a harsh reading of the current academic landscape, and a less-than-generous reading of my decision to dwell on the limitations of (my) research could see it as an attempt to avoid responsibility for my observations and the conclusions I draw from them, but this is not my intention. I could have provided a much more straightforward analysis of the ways in which various actors negotiate climate change and the ways Indigenous Peoples and their identities are constructed through such interactions. I could also have avoided engaging with feminist, decolonial, and post-colonial perspectives altogether, and there are many studies that do so and still provide useful information and important insights (e.g., Brugnach, Craps, & Dewulf, 2017; Larson, 2010; Tsosie, 2007). But to my mind there is little to be gained at this point in history from simply adding to an extensive and ever-growing literature in anthropology and development studies that describes and problematises indigenous struggle (see e.g., Gomes, 2012).[13]

Like Johnson (2008, p. 127), I intend to "free myself from (the myth of) objectivity" by recognising and identifying my own positionality and referring to it throughout: I choose to problematise the researcher and the research, rather than the researched. It is not easy, though, to acknowledge and concede complicity in, while simultaneously attempting to be critical of, the structural violence against Others. Being self-critical does not let the researcher off the hook, nor should it, but being as aware as possible of these limitations and inviting external critique seems, to me, to be the best option available. Acknowledging my position, then, as a white, male, British researcher does not mean that my research escapes all (or any) of the critiques that have been levelled at research like mine done by people like me. As noted by Rigney:

> Historically, the tensions and dilemmas between researchers and Indigenous peoples included issues of representation, power and control. These factors were exemplified in a body of knowledge about Indigenous peoples in what nonindigenous authors called "Aboriginalism'. "Aboriginalism" is defined as "the story about Aborigines told by whites using only white people's imaginations". Aboriginal voices do not contribute to this story, so in Aboriginalism, the Aborigines always become what the white man imagines them to be.
>
> *(Rigney, 2006, p. 34)*

With this in mind, this book is intended not to tell a story about Indigenous Peoples using only white people's imaginations. Rather, it aims to better understand how white people's imaginations have contributed to "making up" (Hacking, 1986) a category of people through the production of a body of knowledge and within the policies, practices, and discourses of Eurocentric institutions, particularly in relation to climate change mitigation. As such, I try to say little about Indigenous Peoples and who or what they are from the perspective *of* Indigenous people themselves. I have some thoughts on this, but it is not the question I seek to address, and I am reluctant to suggest that I might know what they think.

There are a number of reasons for this. Firstly, I do not think that the Indigenous people with whom I spent time were entirely "honest" with me because of the power relationships between us, the influence of outside factors (including money, security, and politics), and because of preconceived ideas, on my part and on theirs, regarding our expectations of each other. Secondly, the Indigenous people I did manage to communicate with quite effectively were a very small number in relation to the total global population of those who might be described, or might self-describe, as Indigenous Peoples. Even this small selected group were mostly "elites" within the indigenous political world, which brings with it many of the issues that are associated with elite representation of any group or community. Thirdly, my positionality (as noted above) meant that, explicitly or implicitly, I conducted my research in a manner that sought answers to predetermined questions (though I did allow these to change over time), and that I have analysed and presented the "data" in ways that reflect my own understandings of how relationships are shaped by asymmetrical distributions of knowledge and power. So, I did not succeed in learning a great deal from the Other and quickly became aware of the epistemological inevitability of bumping into the limits to my understanding. As Todd put it:

> When I think I know, when I think I understand the Other, I am exercising my knowledge over the Other, shrouding the Other in my totality. The Other becomes an object of *my* comprehension, *my* world, *my* narrative, reducing the other to me.
>
> *(Todd, 2003, p. 15)*

As such, the idea of a hyphenated decolonial-researcher is a somewhat paradoxical idea, particularly when research is conducted from the geographical and institutional centre of the history of colonialism (universities in Europe). But the idea that is useful here and that summarises the methodological approach of this book, is that the experience of being exposed to the Other, especially over an extended period of time, brings with it the experience of difference. Through reflecting on this experience, the researcher is also confronted with the limits of their own knowledge and learning that they were not aware of before (Jones & Jenkins, 2008, p. 479). If I have learnt anything at all, it is that I still have a great deal to unlearn.

Data Collection and Interpretation

The empirical chapters of this book are informed by a long-term, multi-sited ethnographic study conducted between March 2014 and January 2016 in Ecuador, Peru, and France. It also includes some reflection on previous time spent in Ecuador during my undergraduate degree from September 2009 to August 2010. Finding a way to conduct such fieldwork ethically is among the greatest methodological challenges faced by researchers (Bernard, 2011, p. 26). However, because throughout this book I consider the ethical issues related to conducting research, I will not focus on all of my ethical concerns of data collection here, and instead will only provide a brief overview of the methods used while conducting fieldwork.

In various locations, ranging from an indigenous community in the Ecuadorian rainforest to a UN Conference in Paris, I conducted approximately 50 semi-structured interviews with people including Indigenous people, activists, representatives of national and international NGOs, government ministries, and IPOs, as well as other interested actors. This was supplemented by innumerable unstructured conversations and interactions with a broad range of people in the different fieldsites mentioned above. Though the interviews are referred to occasionally, the majority of data comes from either participant observation, describing how various moments and interactions appeared to me, or (particularly in the final empirical chapter) reflections on my own experiences while conducting my research. As such, I offer a highly subjective account of the data, but it is, nonetheless, intended to be as honest as my memory allows.

The choice to conduct interviews was based upon the belief that they allow researchers to explore multiple voices and encourage individual experiences and views to be expressed (Byrne, 2004, p. 182). The unstructured, in depth, ethnographic approach to conducting interviews supposedly enables people to be observed in their "natural settings", allowing more nuanced social meanings to be captured (Brewer, 2000, p. 6). Silverman (2005, p. 19) claims that this method of investigation is fundamental to qualitative research because, by conducting interviews in the context of a relationship built through a sustained presence and an established rapport, interviewer and interviewee can explore meanings and understandings of the social world, rather than simply ask and answer a string of questions (Heyl, 2001, p. 370). Unless, of course, you feel like you are not very good at it. Or more to the point, if you do not believe it. The Indigenous people with whom I spent time were not always in their "natural setting" (whatever that means) and the rapport that we had was limited and distorted by our mutual (mis)understandings of the roles we were playing, and of our positions and identities in relation to one another.

That said, there were occasional moments when I felt that an open rapport emerged, and I have drawn upon these where appropriate. But something that troubles me, ethically speaking, is that although I was always honest about my role as a researcher, and always gained oral consent (or written consent when possible and appropriate), I felt as though in the moments when people opened up to me most that they did not *think* they were being interviewed. For this reason, I have made the decision to anonymise all individuals, regardless of having gained consent, other than those who were speaking in a public forum (such as the UN conferences) or had made their names public in interviews elsewhere. The violence encountered by the Sápara during and since the time my research research was conducted has meant that I have also decided to not name the specific community I visited, though the names of COICA and the IPCCA (the main NGOs with whom I interacted) have been kept as it would be impractical to discuss them otherwise.

Participant observation turned out to be the most important tool I had available to me. Though relatively unsystematic when compared with more "refined" research techniques (including interviews, surveys, and quantitative methods), participant observation can at times provide more revealing data (Pelto & Pelto, 1978, pp. 68–69 and 77). In field research, the strength of participant observation is its flexibility; it has neither predefined research method, nor an inflexible, hypothesis. Instead, data collection and sampling techniques depend upon interpretive strategies that allow the "problem" to evolve during research (Crabtree & Miller, 1999, p. 5). Thus, methods such as interviews and document analysis are better viewed as apparatus that *assist* participant observation, and, though research questions are provided, it should be understood that they are not the ones I began with and I don't claim to have answered them in full. My initial hypothesis and questions were tools that guided the research to more refined questions, and these remain only partially answered.

With its flexible and iterative nature in mind, ensuring replicability of participant observation is uniquely difficult, not least because the places and people who participate and observe are changed by the process. Though replication is rare in the social sciences anyway, rigorous documenting of methods, theories, and results that allow research to be repeated is highly valued (Bryman, 2012, p. 47). But as ethnographic research cannot be replicated, attempts to create replicable data limit research to preconceived acceptable practices, sacrificing richness of data (Goetz & Le Compte, 1982, p. 35). Accordingly, textbooks (e.g., Crabtree & Miller, 1999, p. 335) and field research guides (e.g., DeWalt & DeWalt, 2010, p. 128) recommend triangulation of data and methods that emphasise *reliability*, rather than replicability. It would certainly be impossible to replicate any of the data I have produced. One reason for this is that some of the places (and as I argue, the *people*), no longer exist. But another (and possibly more important) reason that this research cannot be reproduced or replicated is that it is largely a subjective account of a personal experience. A more useful way to enter into a dialogue with this research would be to discuss it in relation to other subjective accounts of similar but different experiences in order to identify patterns, commonalities, and shared experiences or highlight anomalies, inconsistencies, and differences. In line with this, this research incorporates a "between method" triangulation (see e.g., Smith, 1978, p. 327), making use of a broad range of existing literature which is referred to alongside discussions of the semi-structured and unstructured interviews conducted.

The best way I have found in the literature to describe the process through which my interviewees and participants were chosen is a respondent driven "snowball" sampling approach, which is understood to be well suited to researching "hard to find" or "hard to reach" populations (Bernard, 2011, p. 192). Though in my research proposal I noted that the eventual aim of this approach was a saturation of the pool of participants, whereby one continues interviewing people until no new names are offered, signifying a completion of the sample (Mason, 2010), the reality was quite different to this. What I actually did was just show up at places and hope that people would speak to me or just let me hang around. Sometimes they did, and sometimes they didn't. Sometimes they barely seemed to notice me at all, and sometimes they decided they never wanted to speak to me again. I chose places where I knew that I would encounter the people *I* was interested in: Indigenous Peoples who were in contact with, or in conflict with, REDD+ or similar projects. But this "choice" of participants was, in a way, an illusion. I do not want to claim that I used a specific method of sampling, because although I tried to select my participants, they and various other factors also selected me.

There is, however, one way in which my position means that my "choice" is the final say, which is in writing this book. What this means is that at times knowingly and at other times not, I have selected those examples from my research that support the argument I am trying to make. The same can be said about the secondary sources I have engaged with and the style in which I have written. Both of these are heavily influenced by factors relating to my identity as a researcher. Though I do at times refer to Indigenous scholars and decolonial and feminist theorists, this book is still influenced by and situated within the predominantly white, male, Eurocentric epistemological community of Western academia, which is fine. It is just important to note that where I have "used" knowledge that might have in the past been situated outside of this community, there is a risk of co-opting and assimilating these ideas into the dominant hegemonic culture. One of my reservations about this is that in doing so, I may have misunderstood or misrepresented the intended meaning of texts, particularly of authors whose work has been translated, but also where I have just missed the point. I hope not, but perhaps in the process of trying to engage in meaningful ways with unfamiliar ideas, this is inevitable. The very act of writing, then, is also an act of taking up space and making up Others.

I am also aware that the choices I have made on terminology including the decision to use the term "indigenous" at all, and sometimes using terms like "we" and "they", brings with it its own set of issues. As noted earlier, what "they" want to be called is a point of contention (Bird, 1999, pp. 1–21), and I am using the term "indigenous" because that is the *idea* that I wish to consider and because it is the dominant term used in relation to climate change. There are four ways in which I have decided to write the word in this book. The first is with lowercase "i" – "indigenous" – which I use when using the term generically and not describing a person or people. When referring to a person or a number of people that identify as Indigenous or that I have *assumed* identify as such but are not representative of *all* people who might fit within the category, I use capital "I" lowercase "p" and the singular "Indigenous people". When referring to the broad category of people who might be identified as, or identify themselves as, Indigenous, I use the uppercase "I" and "P" and the plural form – "Indigenous Peoples" (unless a direct quotation). And finally, I use *Indigenous*, uppercase "I" and italicised, to distinguish a particular way of *being* that I observed among some people who are Indigenous. While there are many differing views on the capitalisation of the term and of words relating to identities in general, and though this approach is imperfect, for the purposes of this book, it allows me to draw some distinction between my intended meanings when using the term.

As for using "we" and "they", I also do this with an awareness of how problematic it is, and that is also my reason for doing. If the reader dislikes this or is uncomfortable with apparently neutral categories such as "white men" or "Western academics" being named, I would encourage them to consider *why* it feels uncomfortable to hear these categories made explicit, particularly if they would consider themself to be part of the category. The point here is that I think there is something intuitively annoying about someone else describing a category you consider yourself part of, and especially so when characteristics assigned to the category are based on broad assumptions. And yet "we" do it to "them" all the time. It is difficult to find examples of research that is not swamped in assumptions of Otherness and Neutrality. As Brons (2015, p. 79) noted "by necessity all interpretation of the other takes place in the terms of the interpreting self, and with that interpreting self as model, at least initially. There is [...] no neutral ground, no interpretation outside schemes", and so I think it is best, at least initially, to face such assumptions head on. Having laid out my approach and broader thinking regarding the form and nature of my research, I begin the empirical part of this book with the first site of ethnographic encounter, which was the final place I visited during my fieldwork: COP21 in Paris, in December 2015.

Notes

1 See, e.g., *What We Want to be Called: Indigenous Peoples' Perspectives on Racial and Ethnic Identity Labels* (Bird, 1999, pp. 1–21). Also, the title of an article by Berglund, Carlson, Harris, & Poata-Smith (2014) shows the extent to which the term "indigenous" has become accepted, because, though it discusses the issues relating to various similar terms, it is itself entitled *Navigating the Complexities of Naming in Indigenous Studies*.

2 This is demonstrated by the ubiquity of the term in climate change debates, not least the common assertion that "indigenous" people have a unique "role" in combatting climate change (see e.g., Etchart, 2017, p. 2; UNPFII, 2007, p. 1).

3 The issue of defining what it means to be indigenous was noted in 1986 by the Special Rapporteur on the Problem of Discrimination against Indigenous Populations, José R. Martínez Cobo, who stated that: "In the thirty-year history of indigenous issues at the United Nations, and the longer history in the ILO on this question, considerable thinking and debate have been devoted to the question of definition of 'indigenous peoples', but no such definition has ever been adopted by any UN-system body" (CBD, 2013, p. 2). This is still the case, and yet UN documents continue to suggest a number of problematic criteria, based on Martínez Cobo's recommendations, for a general definition of "indigenous peoples, communities and nations" which includes "having a historical continuity with pre-invasion and pre-colonial societies that developed on their territories, consider themselves distinct from other sectors of the societies now prevailing on those territories, or parts of them. They form at present non-dominant sectors of society and are determined to preserve, develop and transmit to future generations their ancestral territories, and their ethnic identity" (UN, 2013, pp. 2–3). This means that the very definition of indigenous peoples that precludes the rights afforded to them in the UN declaration, situates them as subordinate and Other, and requires them to wish to "preserve" distinct "ancestral" characteristics.

4 I have not found a reference that says this about the notion of being indigenous, specifically, but Andrews, who recently started the first Black Studies course in Europe at Birmingham, pointed out in a talk I attended at the University of Portsmouth that universities are not simply racist but are the *source* of racism. The *idea* of race and of racism originated in and emanates from within academic circles (Andrews, 2017). In the same sense, the academy is also not just Othering in relation to indigenous people, it is the source of this Otherness, and of the very concept of the indigeneity itself. As Beverley notes (1999, p. 38) "How could such powerful institutions as the university [...] not be implicated in power?".

5 At the present conjuncture, as decolonialism is becoming a more "mainstream" academic perspective, the idea of "unlearning" is quite popular and is often associated with and tends to be credited to Mignolo and Tlostanova (2012), but I am not entirely sure why. The ideas of unlearning and, similarly, "decolonizing the mind" can be traced at least as far back as Fanon (1963, p. 233) and Wa Thiong'o (1981), and Spivak (1990, p. 10) and in feminist stand point theory the notion of unlearning privilege – of men unlearning "manliness" and white people unlearning "whiteliness" – as a way of becoming less well-assimilated members of the dominant group, has also been present for a long time (Bailey, 1998, p. 40; Frye, 2006, p. 126). Allen (1996, p. 256) also notes that, outside of "academic" literature, "Unlearning Not to Speak", a poem by Marge Piercy (1973 cited in Allen,1996), a feminist/activist from the United States, is

demonstrates how, in verbalising lived experiences, knowledge emerges that can have transformative value.

6 A deluge of online rebuttals and critiques from various perspectives followed the article, ranging from comparing the case for colonialism to holocaust denial (Robinson, 2017), to decolonial critiques that pointed out the implicit racism of the piece (Hira, 2017), and even claims that the piece undermined US foreign policy (Khan, 2017). For a discussion on why the article was useful, as it revealed the extant problem of colonialist thinking, see *A Critique of Colonial Rule: A response to Bruce Gilley* (Klein, 2018).

7 For discussion surrounding various issues in the publication and citation process that shape the ways in which knowledge is (re)produced, including English being a de facto global academic language, the practices of knowledge construction at institutional and local levels, and the emergence of Open Access and social media publishing, see Hyland (2016).

8 For a discussion on the problem of the re-emergence of such ideas in contemporary Western academic culture, see Sultana (2018), who argues that the contemporary "false equivalence of academic freedom and free speech [...] in the age of white supremacy, colonial nostalgia, and anti-intellectualism" means that "In the instances when an academic publication or a research project either openly or indirectly promotes violent subjugation of Other peoples, utilizes flawed and dubious argumentation, or does not engage with existing scholarly work with any substance, all academics have the duty to critique and safeguard academic integrity" (Sultana, 2018, p. 250).

9 It is worth noting here that to not criticize due to the accident of one's birth in a place and position of relative privilege is considered by Spivak to be a pernicious position that serves only to sooth the conscience of the self-critical researcher. It is important to take a risk, and the risk is particularly present when critiquing the Other, "something which you used to dominate". And this is why earning the right to speak first requires (and continues to require) doing "homework" (Spivak, 1990, pp. 62–63).

10 There are many examples of this, but a recent one that I found quite revealing is the critically acclaimed *I, Daniel Blake* (Loach, 2016), which depicts the story of a working-class man who in his late 1950s finds himself unemployed and denied state benefits. The issue of the crumbling welfare state and the economic victimization of working-class people is among the most important social issues in Britain today, and the success of the film was significant. But despite many of my friends who I would consider "middle-class" finding the film very moving ("awful... outrageous... terribly sad"), I perceived it as a (re) presentation of a "good bloke": a member of the hardworking and deserving poor, a working-class hero, with whom the middles class can relate.

11 This includes, but is not limited to, the assumptions that being an "insider" means researcher and participant share some sense of identity, place, or belonging with participants (Mullings, 1999), while "outsiders" hold the advantage of being unaffiliated and appearing more objective to the participant and so are approached with less suspicion (Cook & Fonow, 2005). As an outsider in the field, O'Reilly suggested that it is important to maximize the benefits of this role and adopt a persona of being confident and "sort of in charge" but also to allow interviewees to be the expert as this will elicit more open responses (O'Reilly, 2012, p. 148). But as illustrated in the introduction, my fieldsite meant being an outsider carried with connotations *of* expertise. The idea of the "outsider within" is discussed by see also Hill-Collins (1990). This problem is addressed further in later chapters.

12 I personally find it quite strange that anyone would object to this idea. A physicist can acknowledge that their way of studying the world can tell us little about the emergence of post-modernism in art, and an art historian will not often claim they can explain quantum entanglement, but rarely do academics in one field acknowledge that it is possible that *all* fields of academia might still be missing something.

13 In *Activist Research V. Cultural Critique: Indigenous Land Rights and the Contradictions of Politically Engaged Anthropology*, Hale discusses the difficulty associated with conducting research on themes that pertain to political or social movements and notes that: "when we position ourselves in such spaces, we are also inevitably drawn into the compromised conditions of the political process. The resulting contradictions make the research more difficult to carry out, but they also generate insight that otherwise would be impossible to achieve" (Hale, 2006, p. 98).

References

Allen, A. (2017). *The end of progress: Decolonizing the normative foundations of critical theory*. Chichester: Columbia University Press.

Allen, B. J. (1996). Feminist standpoint theory: A black Woman's (Re)view of organizational socialization. *Communication Studies*, *47*(4), 257–271.

Andrews, K. (2017). *The Battle for Black Studies: Decolonising Knowledge is not just 'Academic'*. Presentation for the Citizenship Race and Belonging network at the University of Portsmouth. Portsmouth, UK.

Archer, L., Hutchings, M., & Ross, A. (2005). *Higher education and social class: Issues of exclusion and inclusion*. London: Routledge.

Asad, T. (1973). *Anthropology and the colonial encounter*. Amherst: Prometheus.

Bailey, A. (1998). Locating traitorous identities: Toward a view of privilege-cognizant white character. *Hypatia*, *13*(3), 27–42.

Barthes, R. (2010). *Roland Barthes by Roland Barthes*. (R. Howard, Trans.). Berkley: University of California Press. Retrieved from https://monoskop.org/images/b/b3/Roland_Barthes_by_Roland_Barthes.pdf

Berglund, J., Carlson, B., Harris, M., & Poata-Smith, E. T. A. (2014). Four scholars speak to navigating the complexities of naming in Indigenous studies. *The Australian Journal of Indigenous Education*, *43*(1), 58–72.

Bernard, H. R. (2011). *Research methods in anthropology*. Oxford: Altamira.

Beverley, J. (1999). *Subalternity and representation: Arguments in cultural theory*. London: Duke University Press.

Bhabha, H. K. (1994). *The location of culture*. New York, NY: Routledge.

Bird, M. Y. (1999). What we want to be called: Indigenous Peoples' perspectives on racial and ethnic identity labels. *American Indian Quarterly*, *23*(2), 1–21.

Blanchard, P. (Ed.). (2008). *Human zoos: Science and spectacle in the age of colonial empires*. Liverpool: Liverpool University Press.

Bourdieu, P. (1984). *Homo Academicus*. Stanford: Stanford University Press.

Brewer, J. (2000). *Ethnography*. Buckingham: Open University Press.

Brons, L. L. (2015). Othering, an analysis. *Transcience, a Journal of Global Studies*, *6*(1), 69–90.

Brugnach, M., Craps, M., & Dewulf, A. (2017). Including Indigenous peoples in climate change mitigation: Addressing issues of scale, knowledge and power. *Climatic Change*, *140*(1), 19–32.

Bryman, A. (2012). *Social research methods*. Oxford: Oxford University Press.

Buckle, J. L., & Dwyer, S. C. (2009). The space between: On being an insider-outsider in qualitative research. *International Journal of Qualitative Methods*, *8*(1), 54–63.

Byrne, B. (2004). Qualitative interviewing. In C. Seale (Ed.), *Researching society and culture* (pp. 179–192). London: Sage.

Case, K. A. (2012). Discovering the privilege of Whiteness: White Women's reflections on anti-racist identity and ally behavior. *Journal of Social Issues*, *68*(1), 78–96.

CBD (2013). *Ad Hoc Open-Ended Inter-Sessional Working Group on Article 8(J) and Related Provisions of the Convention on Biological Diversity*. Eighth meeting, Montreal, 7–11 October 2013. Retrieved from https://www.cbd.int/doc/meetings/tk/wscblac-01/information/wscblac-01-inf-01-en.doc

Chen, L. C. (2014). *An introduction to contemporary international law: A policy-oriented perspective*. Oxford: Oxford University Press.

Conklin, B. A., & Graham, L. R. (1995). The shifting middle ground: Amazonian Indians and eco-politics. *American Anthropologist*, *97*(4), 695–710.

Cook, J. A., & Fonow, M. (2005). Feminist methodology: New applications in the academy and public policy. *Signs: Journal of Women in Culture and Society*, *30*(4), 2211–2236.

Crabtree, B. F., & Miller, W. L. (1999). *Doing qualitative research*. London: Sage.

David, A. D., & Wildman, S. M. (1994). Language and silence: Making systems of privilege visible. *Santa Clara Law Review*, *35*(3), 881–906.

Davies, H. (2016). *Neo-Victorian freakery: The cultural afterlife of the Victorian freak show*. London: Springer.

DeWalt, K., & DeWalt, B. R. (2010). *Participant observation: A guide for fieldworkers*. Plymouth: Altamira.

Du Bois, W. E. B. (1903/2008). *The souls of black folk*. Oxford: Oxford University Press.

Escobar, A. (1991). anthropology and the development encounter: The making and marketing of development anthropology. *American Ethnologist*, *18*(4), 658–682.

Etchart, L. E. (2017). The role of indigenous peoples in combating climate change. *Palgrave Communications 3*, 17085. Retrieved from https://www.nature.com/articles/palcomms201785

Fanon, F. (1963/2004). *The wretched of the earth* (R. Philcox, Trans.). New York: Grove Press. (original work published 1963).

Fraser, N. (1981). Foucault on modern power: Empirical insights and normative confusions. *Praxis International*, *1*(3), 272–287.

Frye, M. (2006). White woman feminist. In C. M. Koggel (Ed.), *Moral issues in global perspective II: Human diversity and equality* (pp. 116–128). Plymouth: Broadview Press.

Gal, S. (1998). Multiplicity and contestation among linguistic ideologies. In B. Schieffelin, K. Woolard, & P. V. Kroskrity (Eds.), *Language ideologies: Practice and theory* (pp. 317–331). New York: Oxford University Press.

Gardner, K., & Lewis, D. (1996). *Anthropology, development and the post-modern challenge*. London: Pluto Press.
Goetz, J. P., & Le Compte, M. D. (1982). Problems of reliability and validity in ethnographic research. *Review of Educational Research*, *52*(1), 31–60.
Gomes, A. (2012). Alter-native 'Development': Indigenous forms of social ecology. *Third World Quarterly*, *33*(6), 1059–1073.
Greene, J. P. (2004). The making of a historian: Some autobiographical notes. In J. B. Boles (Ed.), *Shapers of Southern history: Autobiographical reflections* (pp. 18–39). London: University of Georgia Press.
Hacking, I. (1986). Making up people. In D. E. Wellbery, M. Sosna, & T. C. Heller (Eds.), *Reconstructing individualism* (pp. 222–236). Stanford: Stanford University Press.
Hale, C. R. (2006). Activist research V. Cultural critique: Indigenous land rights and the contradictions of politically engaged anthropology. *Cultural Anthropology*, *21*(1), 96–120.
Haraway, D. (1988). Situated knowledges: The science question in feminism and the privilege of partial perspective. *Feminist Studies*, *14*(3), 575–599.
Helland, L. E. F., & Lindgren, T. (2016). What Goes around comes around: From the coloniality of power to the crisis of civilization. *Journal of World-Systems Research*, 22(2), 430–462.
Heyl, B. S. (2001). Ethnographic interviewing. In P. Atkinson, A. Coffey, S. Delamont, J. Lofland, & L. Lofland (Eds.), *Handbook of ethnography* (pp. 369–383). London: Sage.
Hill-Collins, P. (1990). Toward an Afrocentric feminist epistemology. In N. K. Denzin, & Y. S. Lincoln (Eds.), *Turning points in qualitative research: Tying knots in a handkerchief* (pp. 47–73). Oxford: Rowman & Littlefield.
Hira, S. (2017, September 22). A decolonial critique of the racist case for colonialism. *Decolonial International Network*. Retrieved from https://www.din.today/wp-content/uploads/2017/09/The-Racist-case-for-colonialism.pdf
Hyland, K. (2016). *Academic publishing: Issues and challenges in the construction of knowledge-Oxford applied linguistics*. Oxford: Oxford University Press.
Johnson, J. T. (2008). Kitchen table discourse: Negotiating the 'Tricky Ground' of indigenous research. *American Indian Culture and Research Journal*, *32*(3), 127–137.
Johnson-Bailey, J., Kee, Y., Lee, M. Y., Merriam, S. B., Muhamad, M., & Ntseane, G. (2001). Power and positionality: Negotiating Insider/outsider status within and across cultures. *International Journal of Lifelong Education*, *20*(5), 405–416.
Jones, A., & Jenkins, K. (2008). Rethinking collaboration: Working the indigene-colonizer hyphen. In N. K. Denzin, Y. S. Lincoln, & L. T. Smith (Eds.), *Handbook of critical and indigenous methodologies* (pp. 471–486). Thousand Oaks: Sage.
Kerstetter, K. (2012). Insider, outsider, or somewhere in between: The impact of researchers' identities on the community-based research process. *Journal of Rural Social Sciences*, 27(2), 99–117.
Khan, S. (2017, September 19). The Case Against "The Case for Colonialism". *Cato Institute*. Retrieved from https://www.cato.org/publications/commentary/case-against-case-colonialism

Klein, M. A. (2018). A critique of colonial rule: A response to Bruce Gilley. *Australasian Review of African Studies*, *39*(1), 39–52.
Kuper, A., Omura, K., Plaice, E., Ramos, A., Robins, S., & Suzman, J. (2003). The return of the native. *Current Anthropology*, *44*(3), 389–402.
Larson, A. M. (2010). Making the 'Rules of the Game': Constituting territory and authority in Nicaragua's indigenous communities. *Land Use Policy*, *27*(4), 1143–1152.
Lévi-Strauss, C. (1966). The scope of anthropology. *Current Anthropology*, *7*(2), 112–123.
Lins Ribeiro, G. (2016). Outsiders and insiders in the making of anthropological knowledge. *American Anthropologist*, *118*(3), 628–629.
Loach, C. (Director). (2016). *I, Daniel Blake*. United Kingdom: Sixteen Films/British Film Institute/Why Not Productions/Wild Bunch/BBC Films
Mason, M. (2010). Sample size and saturation in PhD studies using qualitative interviews. *Forum: Qualitative Social Research*, *11*(3). Article B. Retrieved from http://www.qualitative-research.net/index.php/fqs/article/view/1428
May, T. (2016). *Theresa May's first speech as Prime Minister: full text*. Retrieved from https://www.spectator.co.uk/article/theresa-may-s-first-speech-as-prime-minister-full-text/
McDowell, L. (2011). *Redundant masculinities: Employment change and white working class youth*. London: Blackwell.
Mignolo, W., & Tiostanova, M. V. (2012). *Learning to unlearn: Decolonial reflections from Eurasia and the Americas*. Columbus: Ohio State University Press.
Morton, P. A. (2003). *Hybrid modernities: Architecture and representation at the 1931 colonial exposition, Paris*. Cambridge: MIT Press.
Mullings, B. (1999). Insider or outsider, both or neither: Some dilemmas of interviewing in a cross-cultural setting. *Geoforum*, *30*(4), 337–350.
O'Reilly, K. (2012). *Ethnographic methods*. London: Routledge.
Pels, D. (2000). Reflexivity: One step up. *Theory, Culture & Society*, *17*(3), 1–25.
Pelto, G. H., & Pelto, P. J. (1978). *Anthropological research: The structure of inquiry*. London: Cambridge University Press.
Ploder, A., & Stadlbauer, J. (2016). Strong reflexivity and its critics: Responses to autoethnography in the German-speaking cultural and social sciences. *Qualitative Inquiry*, *22*(9), 753–765.
Presser, L. (2005). Negotiating power and narrative in research: Implications for feminist methodology. Signs: *Journal of women in culture and society*, 30(4), 2067–2090.
Quijano, A. (2000). Coloniality of power and Eurocentrism in Latin America. *International Sociology*, *15*(2), 215–232.
Rabe, M. (2003). Revisiting 'insiders' and 'outsiders' as social researchers. *African Sociological Review/Revue Africaine de Sociologie*, *7*(2), 149–161.
Radcliffe, S. A. (2015). *Dilemmas of difference: Indigenous women and the limits of postcolonial development policy*. Durham: Duke University Press.
Riessman, C. K. (2015). Entering the hall of mirrors: Reflexivity and narrative research. In A. De Fina, & A. Georgakopoulou (Eds.), *The handbook of narrative analysis* (pp. 219–239). Chichester: Wiley.

Rigney, L. I. (2006). Indigenist research in aboriginal Australia. In J. E. Kunnie, & N. I. Goduk (Eds.), *Indigenous Peoples' wisdom and power: Affirming our knowledge through narratives* (pp. 32–51). Aldershot: Ashgate.

Robinson, N. J. (2017, September 14). A Quick Reminder of Why Colonialism was Bad. *Current Affairs*. Retrieved from https://www.currentaffairs.org/2017/09/a-quick-reminder-of-why-colonialism-was-bad

Ruth, D. (2008). Being an academic: Authorship, authenticity and authority. *London Review of Education*, 6(2), 99–109.

Sachs, W. (1993). *The development dictionary: A guide to knowledge and power*. London: Zed Books.

Sillitoe, P. (Ed.). (2016). *studies and engaged anthropology: The collaborative moment*. London: Routledge.

Silverman, D. (2005). *Doing qualitative research: A practical handbook*. London: Sage.

Smith, L. M. (1978). An evolving logic of participant observation, educational ethnography, and other case studies. *Review of Research in Education*, 6, 316–377.

Smith, L. T. (1999). *Decolonizing methodologies: Research and indigenous peoples*. London: Zed Books.

Spivak, G. C. (1990). *The post-colonial critic: Interviews, strategies, dialogues*. London: Routledge.

Sultana, F. (2018). The false equivalence of academic freedom and free speech. *ACME: An International Journal for Critical Geographies*, *17*(2), 228–257.

Tascón, S. (2004). Refugees and the coloniality of power: Border-crossers of postcolonial Whiteness. In A. Moreton-Robinson (Ed.), *Whitening race: Essays in social and cultural criticism* (pp. 239–253). Canberra: Aboriginal Studies Press.

Taylor & Francis (2018). Withdrawal Notice. *Third World Quarterly*. Retrieved from https://www.tandfonline.com/action/showCitFormats?doi=10.1080%2F01436597.2017.1369037

Todd, S. (2003). *Learning from the other: Levinas, psychoanalysis, and ethical possibilities in education*. New York: State University of New York Press.

Trupp, A. (2011). Exhibiting the 'Other' then and now: 'Human Zoos' in Southern China and Thailand. *Austrian Journal of South-East Asian Studies*, 4(1), 139–149.

Tsosie, R. (2007). Indigenous People and Environmental Justice: The Impact of Climate Change. *University of Colorado Law Review*, 78, 1625–1677.

UN (2013). *Fact sheet 9/Rev. 2- Indigenous peoples and the United Nations human rights system*. New York & Geneva: United Nations. Retrieved from https://www.ohchr.org/Documents/Publications/fs9Rev.2.pdf

UNPFII (2007). Indigenous Peoples, Indigenous Voices: Climate Change and Indigenous Peoples. *Backgrounder: Indigenous Day*. Retrieved from http://www.un.org/en/events/indigenousday/pdf/Backgrounder_ClimateChange_FINAL.pdf

Wa Thiong'o, N. (1981). *Decolonising the mind: The politics of language in African literature*. London: James Currey.

Walsh, C. (2007). Shifting the geopolitics of critical knowledge: Decolonial thought and cultural studies 'Others' in the *Andes. Cultural Studies*, *21*(2–3), 224–239.

Wasserfall, R. (1993). Reflexivity, feminism and difference. *Qualitative Sociology*, *16*(1), 23–41.

Whiffen, M. (2004). *The Anthropologist as Apologist? Colonial Heritage, Ethnographic Representation and the Ethics of Advocating Practice*. ADST MA Programme 2003-04. Retrieved from https://www.sussex.ac.uk/webteam/gateway/file.php?name=whiffen&site=10

PART II

Seeing and Being Indigenous

PART II

Social and Health Inequalities

3

NEGOTIATING INDIGENEITY AT THE INDIGENOUS PAVILION

Introduction

In Le Bourget, in the North-Eastern suburbs of Paris, on November 30, 2015, 12 km or so from the centre of the city, the second International Indigenous Peoples Pavilion appeared. It was nestled in a corner of the Climate Generations Space at COP21 and on December 12th, like the first Pavilion had a year before, it would promptly disappear. It would not go somewhere else but would simply cease to exist. Its constituent parts – the structures and the chairs, the fixtures and fittings, and the pictures and artefacts on display – may or may not still exist somewhere, but the place itself, the Indigenous Pavilion, does not. Likewise, the Indigenous people for whom and by whom it was created still exist somewhere, but the people that they collectively became whilst there are, like the place, also gone. Though the physical space and people who attend might be different, the Pavilion will later be recreated and reappear in Marrakech for COP22 in 2016, then in Bonn for COP23 in 2017, and so on in various locations for years to come.

The Indigenous Pavilion is one of the many spaces that emerge on cultural frontiers where eco-politics meets indigenous struggle; it is a "middle ground", or "mutually comprehensible" space (Conklin & Graham, 1995) where understandings of ideas, identities, and environmental concerns that sometimes appear contradictory are negotiated and (re) created. It is also a place where some people who are Indigenous apparently become something else. They embody the "biosocial fact" (Hartigan, 2013, pp. 4 and 17) of being indigenous *to* somewhere but also

DOI: 10.4324/9781003341864-6

take on and project a complex web of myths and fantasies (see e.g., Paradies, 2006; Salazar, 2013) associated with and encompassing indigenous identity, though these characteristics may have little to do with being indigenous *to* anywhere in particular. Sometimes, then, there is an apparent difference between being Indigenous and being *Indigenous*. Though constructing and "proving" indigeneity can be an effective strategic tool in these types of negotiations, it also carries with it risks and contradictions that could potentially undermine the political agenda of Indigenous Peoples (Conklin, 1997; Tsing, 2007), resulting in what Baker, Brock, Dove, Hebdon, and Ludlow (2016, p. 2) describe as the "double bind": a contradiction whereby, when deploying the concept of indigeneity, "the more modern or global [...indigeneity...] is seen as being, the more its authenticity as an identity is questioned". Li (2000, p. 12) acknowledged that "The idea that there is an epistemological or substantive distinction between indigenous knowledge and other kinds of knowledge (Western, scientific, non-indigenous) has been, quite rightly, debunked" by Agrawal (1995, cited in Li, 2000, p. 12). And yet the distinct category still exists and has been afforded an elevated status in political discussions around climate change. So, questioning, or at least attempting to understand, who the *Indigenous* people in these negotiations are and what indigeneity means in this context, is a good place to start when seeking to better understand who is legitimising whom when Indigenous Peoples are present in non-indigenous spaces where environmental issues are being debated.

I present here a selection of vignettes; snapshots from my observations and interactions in and around the Pavilion at COP21 in Paris. This was where, after the thoughts began to emerge a year earlier at COP20 in Lima, the idea of being *Indigenous* came into focus and I began seeing it as related to, but distinct from, being Indigenous. The intention of this chapter is to illustrate some of the ways in which people are being and becoming something unique at the conference, as well as to reflect on the sometimes-uncomfortable position of being a researcher who goes into the field with the intention of seeing *Indigeneity*. I discuss this distinct category of *Indigenous* people, arguing that it is made up both by and for the context in which it emerges. As such, it is also, like the middle ground in which it appears, a transient phenomenon; it exists only for as long as the context itself. *Indigenous* people, then, can appear, disappear, and reappear quite suddenly: sometimes a person who is Indigenous is also *Indigenous*, and sometimes they are not. It can all be somewhat disorientating for a non-indigenous person, especially when trying to understand what indigeneity means, how it relates to being *Indigenous*, and why this matters in global environmental negotiations.

The Canoe

In the afternoon of a cold but sunny day, a young man stands in the space between the two main structures of the Indigenous Pavilion in the Climate Generations Space of COP21: a vast white temporary marquee filled with stalls, exhibits, conference spaces, and other pavilions. It is about as far from the entrance as you can get, but its position alongside the food vendors guarantees a fair amount of foot traffic. To the right of the young man stands the larger of the two structures: a rectangular room with green carpets and green walls, and oversized pot plants in the corners. Rows of chairs, around 100 seats in total, face the array of televisions that cover much of the front wall. The screens are the backdrop to a small, raised platform, which is flanked by more plants, and to one side there is a lectern. Like the walls, it is made of recycled chipboard and on the front is the Indigenous Pavilion logo, a green circle with a chain of seven white hands encircling an impression of the sea, mountains, the sun, and the sky. There are recorded sounds of birds coming from somewhere. The parts of the walls that are not green are decorated with images of Indigenous Peoples, doing indigenous things, in indigenous places.

To the man's left is the smaller structure, a facsimile of a generic *Indigenous* building. Its eight sides, four of which are open, and the low-pitched roof are somewhat hoganesque in appearance. I sat in the building early one morning before the panels began and read about the cultural significance of this type of structure. The "*hooghan*" (anglicised to hogan) is a Navajo building, and this one at the Indigenous Pavilion resembled the female kind, the *hooghan namazi*, which is usually made of logs and mud, with a peaked roof and six or eight low walls. The hogan is a traditional home and a communal space, but it is also, according to Lane (1999, p. 37), part of a situated knowledge; it embodies social, philosophical, and spiritual teachings of the Navajo people. It is part of the creation story, and more than just a place to sleep and eat, it functions as a temple and an animate "being", which must also be fed and sustained (Callaway et al., 1968, p. 56, cited in Manuelito, 2005, p. 81). But this one, in Paris, is made of panels of chipboard and semi-opaque plastic. It is decorated in a similar way to the other room, but with an earthy orange carpet that evokes (intentionally or not) the dry sandy earth of a desert rather than a forest floor. The chairs in this room are arranged differently, facing the centre instead of the front, reflecting the octagonal structure and creating a more communal space.

The young man standing between the structures is clearly *Indigenous* because of his brown skin, long, thick, black hair, and the geometric lines painted on his face.[1] He is speaking with another man. This one is a little older (in his 30s), and he is clearly *not* Indigenous because his hair is short and thin (and a little grey), his skin is pale, and he does not wear makeup. He does not need to. His identity is not intended to be part of a group: he thinks of himself as an observer and as an individual. They stand alongside a dugout canoe: 30-foot-long, with carvings of serpents along the sides. The second man is interested in the first because he is *Indigenous*, and so he asks:

> "Why was it important for you to bring the Canoe here?"

> "The Canoe is to bring a message of *Kawsak Sacha*, Living Forest, from our people" responds the *Indigenous* man.

As he speaks beads of sweat begin to form on his top lip and above his eyebrows. He seems uncomfortable; maybe nervous or shy. He appears to be panicking.

"And what does the Canoe signify for you?" The second man asks.

After a brief pause, the Indigenous man say:

"The Canoe, for us… is used for transport by our people."

Apparently, this is not the response the second man is looking for. He rephrases the question to make it more leading and pushes for a "better" answer; maybe something... more profound, more symbolic or poetic. It will sound better when he recounts the story in his thesis.

> Are there any… other reasons why it is important? Is it *only* for transport, or is it significant for any other reason… for you [all]?

He asks the question without realising at the time that by switching from the Spanish singular to the plural form of "you" (*ustedes*), he is implicitly asking the man to speak for *all* Indigenous Peoples. No wonder he is nervous. After a short pause, the first man responds once more:

> The Canoe has a long history. It has been around as long as the moon, and we continue to use it up to today

Unfortunately, the *Indigenous* man, Esteban, does not know his lines as well as the non-indigenous man had hoped, but this will have to do. The non-indigenous man already knew the answer that he wanted. He had read and heard much about the role and importance of the canoe in indigenous cultures and knew why this particular one was at COP21. Esteban could have said, the white man thought, that:

> It is used for transport, yes, but more than that it connects people and communities, creating networks and social ties that allow us to share with and learn from one another.[2] From outside, it brings us food and other resources and brings visitors into our world. It takes parts of our culture, the things that we create, and carries them and *us* to the outside.[3] In our stories of creation and the myths of gods it is a common symbol: it carries carries them and the spirits of our ancestors to us from other worlds, bringing us new knowledge and wisdom.[4]

> He *should* have said:

> It too has a spirit.[5] It is crafted by many of us, from a forest tree; it *is* the forest[6]. It *is* indigenous…

> Or something like that.

The conversation does not get this far. The white man is gently elbowed aside by another white man who wants to take a picture of Esteban. It is taken without asking, but Esteban does not seem to mind. He seems relieved to be away from the loaded questions. He turns away, wipes the sweat from his eyes and mouth. His makeup smudges a little, and he looks back towards the photographer and he smiles. Another picture is taken. The photographer does a half bow with his hands held together as if he is praying.

> "Thank you… gracias… thank you" he says, with a smile and an American accent.

Observing this interaction between the photographer and Esteban forced me to reflect once again on what I was doing with the questions I was asking. Like the photographer, my primary concern was to take something, and the closer it was to my existing notions of what it is to be "indigenous" the better. I should have known better, though. I had read the critical work of Indigenous scholars on decolonising methodologies (see e.g., Denzin, Lincoln, & Smith, 2008), but the fundamental problem remained: "the researcher's assumption of a 'right to know' that serves as a carte blanche for activities that contradict other values" (Reinharz, 1998, p. 95).[7]

Esteban has had these interactions many times today and on many other days. Outsiders have visited his community since he was a child:

reporters, researchers, NGO workers, scientists, and government officials have all come and gone. Even more so since his people won a landmark lawsuit against the Ecuadorian Government in the Inter-American Court of Human Rights for violations of their rights during the expansion of oil exploration into their territory. He is used to being *Indigenous*. He is also used to cameras. Esteban is himself a documentary filmmaker.[8]

It is the last day of COP21. The Canoe for Life was dugout and carved by hand months ago, 10,000 km away in an indigenous territory deep in the Ecuadorian rainforest. It travelled along tributaries of the Amazon River, went overland to the coast, and was shipped across the ocean to France. It was then paddled ceremoniously along the Canal de la Villete in the heart of Paris, and finally, it was carried into the Indigenous Pavilion on the shoulders of men. The men who carried the Canoe were Indigenous. They were also *Indigenous* because they were carrying the canoe, like the men earlier in the week who had been *Indigenous* when they carried a block of ice through the Climate Generations Space, beating drums and chanting, conducting a symbolic funeral for the melting ice caps. Or like the man from Vanuatu, who came to the Pavilion with kava.[9] He scooped it from an ornate bowl and shared it with Indigenous and non-indigenous people from around the world. And the women from the Amazon had been *Indigenous* when they danced in the Pavilion, swinging their long black hair while people watched, took photos, and clapped their hands to the beat of a drum. It is all very entertaining and very symbolic. It is very *Indigenous*.

The Canoe for Life is a fitting symbol for indigenous participation at COP. It can be seen as a physical representation of the story of indigenous negotiation at the Conference. It takes on a dual meaning, the first of which relates to the notion of negotiation as an emergent phenomenon, engendered by the intersubjective relationship between actors. According to Duranti's reading of Husserl, intersubjectivity is not simply reaching a shared or mutual understanding of an objective reality as it has often been treated in anthropology but is instead itself the *source* of objectivity. In this sense, intersubjectivity arises when different actors agree that, were they able to "trade places", the world itself would remain the same. In Duranti's words: "the world as it presents itself to me is the same world as it presents itself to you, not because you can 'read my mind' but because I assume that *if you were* in my place you *would* see it the way I see it" (2010, p. 21).

Negotiation between actors is conditioned on this intersubjective agreement, which in turn relies on symbolic markers. The Canoe, like the Pavilion and the Indigenous people in it, is one such symbol to which what it is to be *Indigenous* can be anchored[10] – not because it represents the same thing to all actors, but because all actors agree that, were they to trade places with one another, being Indigenous might mean something different, but the Canoe would still be *Indigenous*. The "Self" (indigenous or otherwise), according to Duranti (2010, p. 23), also depends on the physical presence of an Other, or as Viveiros de Castro (2011, p. 73) puts it, "the absolute necessity of an exterior relation" the absence of which – a world without Others – would be unthinkable. The Canoe, along with other symbols and markers at the Pavilion, helps to delimit and distinguish the negotiated, intersubjective category of *Indigenous* and, in turn, non-indigenous. Though the subjective categorical schemes may be different, the the objective category itself is (to some extent) (to some extent) agreed upon. This categorisation, though it undoubtedly positions Indigenous people as Other in this context (*Indigenous* being almost synonymous *with* Other), also provides a possibility of a shared, intersubjective identity among Indigenous people, which forms the foundation of a collective identity and, in turn, the potential for collective demand-making at COP.

The Shaman and the Salesman

Manari is a Shaman. He takes the lectern in the corner of the large forest themed room in the Indigenous Pavilion. He is either more or less authentically *Indigenous* than Esteban because he looks different. His clothing is more "traditional", his headdress is among the most beautiful and ornate of the many on display at the Pavilion, and his makeup is bolder. He is older, too. He looks strong and stoic: a better symbol of indigenous wisdom, full of knowledge of the forest. The pictures around the room reflect this idea: Indigenous people are either young and innocent or old and wise, and always only a little savage. The images show them fishing in dugout canoes, or dressed in furs, smiling, holding their catch. The pictures do not show them using a chainsaw to cut the tree for a canoe or killing seals for their fur. The image of *Indigenous* people is carefully curated, here.

The awkwardly large pot plants make up a faux forest that sits behind the lectern, and as he speaks the leaves of occasionally catch the feathers on top of Manari's head. He notices but ignores it, as do the audience. They watch and listen respectfully and in silence. He repeats, almost word for word, what he had said in an interview earlier that day. Unlike Esteban, his lines are well rehearsed: his message is clear and concise. He does not appear nervous, but instead he seems calm, and he speaks with intensity and passion:

> Money... for us, money is not the solution. We can win land titles, we can secure reservations and recognition of our rights, but this also is not

> the solution. The solution has to be where the big so-called-developed-countries, the materialist world, recognizes that the problem is with you.

Esteban is filming from the opposite corner of the room, but nobody is filming *him*, now, because he is not being *Indigenous*.

Manari places a plastic bottle on the lectern for the audience to see. It contains a red-brown earthy looking liquid, and the pattern of lines and dots that is painted on his face is precisely recreated on the label. The name of the drink, "Naku", is printed in a font that mimics this same pattern. The branding is subtle, but it is well defined and consistent, repeated again along the length of his ceremonial tunic, made from the bark of a tree, and once more on the business cards he gave me earlier that day. He talks of the healing power of the drink and of the importance of the medicinal knowledge that his people have. He holds up a wrap of tobacco, cured in the traditional manner and bound in twine, and he explains its spiritual healing and cleansing power. You can now go, he says, to his community to visit and to learn about this, and to experience the tranquillity of the forest. Earlier, after the interview, he had asked me, "Did you know Channing Tatum has been?". Manari's website states that "Even if we disappear, we want to leave our knowledge for the world". He wants this knowledge, along with the medicines of the forest, to survive "so that the rest of the world can be cured" and asks "what if we could cure chronic illness and help save the rainforest at the same time?" (Naku, n.d.). Perhaps a week-long trip to the ecotourism project in his territory, at a price of US$3000, can help answer this question (NakuNorth, n.d.).

Manari has positioned himself here, between two worlds that seem to contradict one another. He is both a Shaman and a salesman. His ability and willingness to hold two apparently conflicting ideas in his mind is evident when he speaks. While he criticises the capitalist ideology, the neoliberalism, and materialism of the "so-called-developed countries", he is also able to sell elements of his culture to those that he blames for the destruction of it in order to preserve it and to share it with them. And yet this could be construed as demonstrating an inconsistency or a lack of authenticity, which in turn could undermine his position and his power. As Carlson, Harris, and Poata-Smith (2013, p. 1) have argued, to question indigenous authenticity[11] and to create and imbue such rigid ideas of what is *really* indigenous, with seemingly impermeable boundaries and indestructible meanings, are rooted in colonial racism, which has facilitated the alienation of indigenous land and resources, and which has in turn justified state policies that regulate the daily lives of Indigenous people. Viatori (2005, p. 106) describes the way that Bartolo (Manari's Spanish language name, which I have never known him to use) has presented himself to non-indigenous people on repeated trips to the United States and Europe: "he considers his responsibility to be the representation of the Zápara as authentic Indians to outside audiences [...] Bartolo creatively combines Zápara, Kichwa, and 'traditional' clothes to symbolize his authenticity as an indigenous leader". I remember thinking it must be a difficult balance to strike, to make sure you are perceived as authentic, but without going too far, and later I wondered how it must feel to have to do this to be heard.

Perceived (in)authenticity, though, says little about the internal logic of taking such an apparently contradictory position or to put on what seems to be an "act" or "performance" in order to be *more* authentic. The difficulty non-indigenous people have in consolidating this changeable character of Indigenous people can be traced back to early anthropological encounters, including with the Sápara.[12] Simson (cited in Taussig, 1987) in 1885 considered why the "really wild Indians" were more difficult to conquer and colonise than the more "docile" groups, describing the "Zaparos" as:

> ... changeable and unreliable, betraying under different circumstances, and often apparently under the same, in common with so many of their class, all the opposite traits of character, excepting perhaps servility- a true characteristic of the old world- which I have never observed in them. The absence of servility is typical of all the independent Indians of Ecuador.
>
> *(Simson, cited in Taussig, 1987, pp. 90–91)*

In this sense, the inability (or unwillingness) of Indigenous people to consistently fit into the category of *Indigenous* subverts not just the attempt at categorisation but the categorical scheme itself. It is disquieting because if, as Viveiros de Castro noted, the world without an Other is unthinkable, the world without a consistent Other must be as incomprehensible, or at least as inconstant *as* the Other.

Manari's self-identification as a Shaman may, in this instance, serve him well. To me, the changeability described above, as well as the "inconstancy" that is found in colonial accounts of encounters with Amazonian Indigenous people (and is praised by Viveiros de Castro, 2011, p. 100), are aligned with notions of metamorphosis and shape shifting that are a ubiquitous part of (non-indigenous) descriptions of shamanism in Ecuador, across the Amazon, and elsewhere in the world (see e.g., Praet, 2005, 2009). Shamanism often requires a person to become separate to and different from their community. As Cepek (2012, pp. 90–93) explains with reference to the Cofán people of Northern Ecuador, "A shaman wins his positive powers [...] by identifying himself with all that is violent and morally condemned [...] he speaks in its language and controls the entrance and exit to its house". Though the Shaman may at times be viewed with mistrust, Cepek continues, it is this experience of other worlds and the ability to speak in the language of dangerous Others that is called upon by the community when an alien presence causes social disruption. In this sense, to be a Shaman is to be hyper-indigenous, but paradoxically it also allows for the uptake of typically non-indigenous characteristics in order to negotiate with, but not to serve, outsiders: embodying Otherness while preserving the Self. Manari's combination of pragmatism and "authenticity" appeared, to me, to be both more and less *Indigenous*.

Manari's familiarity with the alien languages of conservation and the "value" of rainforests had been called upon during the negotiation of his people's territory in relation to Socio Bosque, the Ecuadorian national REDD+ readiness programme. Along with two other leaders, he signed (in his Spanish name, Bartolo) the agreement to a government project that would supposedly protect his people and his territory from the risks of oil exploration and deforestation and would also pay the Sápara people for it. He became entangled in a contract that, along with continued expansion of the oil frontier into their territory, would lead to conflict and division within his family and community (Duhalde & Vallejo, 2017, p. 258).

Forgotten Faces

Between the talks at the Indigenous Pavilion, as is often the case at conferences, some people disappear. They go outside to smoke. On a short fenced-in walkway that leads to a separate building where the toilets are housed, a man and a woman stand outside a row of doors near the Pavilion. Because it is the only place in the Climate Generations Space where you can get outside without having to go through the airport-style security checks, which could take up to an hour, it had become the de facto smoking area. The security personnel had initially tried to prevent this, but they had given up after the first few days. With the series of coordinated terrorist attacks by ISIL extremists that had happened in Paris only two weeks before, there were more pressing security issues to deal with. In the first days of the conference, there was a nervous tension among the people attending. The French authorities had placed a ban on outdoor protests and events, and many people had decided not to travel to Paris at all. But quickly the armed security on the metro and the trains began to feel mundane, as did the excessive security controls at the conference site. It felt almost normal, and people seemed more frustrated by the restrictions and inconveniences than they were afraid of another attack.

The woman walks over to the man and, with a big smile, asks him for cigarette. He opens a new pack and hands her two, "There's nowhere on site to buy them", he says. They begin to make small talk about how cold Paris is. She speaks English well but with an accent that the man finds difficult to place. It might be French. She makes a joke about being an endangered species; "Smokers?" he asks, "and Indigenous people" she replies. Then she asks the man to tell her more about his PhD research. He realises that they have met before, and he suddenly feels awkward and anxious. First because he did not recognise her and then because he realises he is going to have to speak to an Indigenous person, again, about what he "knows" about Indigenous people. He had taken a picture of her no more than an hour ago before he knew her name, while she was singing and dancing near the Pavilion. She had been wearing the symbolic markers of her indigenous culture and identity: the boots, the headband, the brightly coloured clothing and jewellery, and she was playing a drum. She is from a hunter culture, indigenous to Siberia: she had told him this earlier when they had spoken briefly before she went on stage. She had asked him why *he* was there, too, and he had not said anything about *his* culture or origins, only that he was doing research about Indigenous Peoples and climate change. He later learned that the drum and the clothing she wore on stage were part of a shamanic ritual:

> ... the shamanic costume itself, which consists of boots (*böö gutal*), gown (*böö malgai*), headgear (*böö malgai*), drum (*hengereg/hets*), and drumstick (*tsohiur/orov*), is usually locked away in drawers beneath the altar, only to be taken out in the final hours before the ceremony. This is not surprising, for in Caroline Humphrey's apt phrasing, the costume is believed to place the shaman in a "world conquering time-machine".
> *(Pedersen, 2011, p. 159)*

These shamanic objects (not unlike the canoe or the hogan) serve as physical containers, receptacles, or vessels for spirits known as *ongod* (Pedersen & Willerslev, 2012, p. 481), and during the shamanic ceremony, the woman is, according to Pedersen (2011, pp. 161–162), possessed by her own *ongod*, a moment which is marked by the drumbeat becoming faster and faster. The Shaman remains possessed until the drumstick is thrown down, and she rushes to remove the clothing because it is considered to be dangerous to wear it while not drumming. It is only then that she becomes herself again.

Out in the cold, smoking a cigarette in Paris, she wears a thick, black coat, no gown, no fur, no hide boots, no headband, and no jewellery, and she is not beating a drum or singing. She is still Indigenous, but she is not being *Indigenous*, in any perceptible sense, by the standards of the conference. The man confesses that he had not realised who she was, and she laughs and reassures him "I look really different in my costume". Perhaps, considering the role of these objects in the shamanic ceremony, the containers of *ongod*, that allow a woman to be possessed, to *be* something else, there was no need to be embarrassed for not recognising her.

*

Back in the Pavilion, the Canoe for Life is encircled by five or six women. They are clearly marked out as *Indigenous* by their short stature and brown skin, their long black hair, and their brightly coloured clothes. The women and the Canoe are being milled around by taller, whiter people, with cameras, notepads, phones, and voice recorders. Some are journalists, some work for environmental NGOs or IPOs that are represented at the Pavilion, some are from other Pavilions and stands in the Climate Generations Space, and some have stumbled upon something interesting; a curiosity, on their way to get food or to use the toilets, or to sneak out for a cigarette. They add their mark to the mural on the wall. They smile and nod at the *Indigenous* women. They photograph them and they admire the Canoe and the small collections of ceramics, jewelry and other artefacts, which are on display on shelves either side of it. On the seats of the Canoe, there are now pieces of colourful bead jewellery, similar to that which is worn by some of the *Indigenous* people, but smaller and more generic. The Canoe became, in a more literal sense, a vessel or container of indigeneity, and its symbolic role in facilitating trade and building networks with Others was reified.

"For your wife?"

One of the women says as she offers a bracelet to a white man who is standing nearby with his phone in his hand and a camera hanging over his shoulder. He was posting a picture from the event Instagram.

"No, but thank you, Ena." He responds in Spanish, using her name.

She is a Sarayaku woman, from the same place as the Canoe. The woman looks surprised and smiles.

"For your daughter?"

"I don't have a daughter but thank you."

"But it is so pretty!"

"Yes, it is. But no, thank you."

Ena, polite but persistent, did not seem to recognise the white man she was selling to, though they had met before. Once or twice in Quito, once after she spoke at an Indigenous women's conference in Puyo (a small rainforest town) a year or so before, and a few days ago at a press conference on a river boat in the centre of Paris. She may have been at COP20 the year before, too. It is hard to keep track. It is always the same people at these things, but it is easy to forget who you have encountered and when. She is here to be *Indigenous*, and he is here to see *Indigenous*. He buys nothing, he never does. Instead, he takes her picture again, takes some notes, thanks her, and then walks away.

Like Esteban, Ena seems quite comfortable with the interaction. All of this has happened before. Indeed, her face is so memorable to the white man because he has seen it so many times. At COP20 it was part of a photographic exhibition of Indigenous women by an Ecuadorian photojournalist: "Amazonas: Guardians of Life". He had also seen the exhibit at a university in Quito and had stumbled across it again when it was reproduced online and in print, including in the Ecuadorian environmental magazine *Terra Incognita*. He bought a copy of the magazine before he left Quito, and it happened to be in his camera bag as he spoke with Ena. Coincidentally, he had travelled to the rainforest a year earlier with Felipe Jacome, the photographer who took her portrait, They had flown together to Sápara territory, to one of the seven communities where the pictures were taken, along with another photographer and two representatives of an environmental NGO, all pursuing their own projects. Each photo in Jacome's collection had been accompanied by passages handwritten by the subject of the portrait in their own language. Ena's said:

> I'm from the community of Sarayacu. I'm a woman who has fought against oil exploitation. In 2002, an oil company came into Sarayacu territory to destroy the Amazon jungle. The women in this community worked side by side with our children, youth, adults, and elders to resist. Even our schools shut down during our struggle against the oil company.
>
> We formed groups and divided the tasks to be able to hold out in our struggle. The women mostly prepared the *chicha* and the food for the men who were out in the jungle. The women of Sarayacu have the dream of continuing to fight to preserve our territory, our jungle, our river, and our air free of pollution. We also fight against violence against women in the community. Let's stand up. It's time to open our eyes. It's time to come together in one heart, pure and strong. It's time we rise again.
>
> *(Santi, 2014, cited in Jacome, 2015)*

In the networks that have developed around the theme of indigenous environmentalism, forgetting people, and being forgotten, is a common occurrence, but it is a bigger mistake for some than it is for others. Though many of the same people rotate around places and events like the Indigenous Pavilion and COP from year to year, and though many of those same people come into contact with one another in other less "formal" spaces, including in indigenous territories, climate change protests, and other similar events, there is often a transient nature to the relationships between Indigenous and non-indigenous people.

During COP21, I met a group of three Navajo men, one of whom I had briefly met a year earlier in Lima. We had had a short conversation about REDD+, lasting only around two or three minutes. When I was reintroduced to them, I got two of their names mixed up and, over the next few days, I continued to refer to them incorrectly until one said, quite abruptly "My name's not Sam!". I was immediately embarrassed and began to apologise. I felt my face beginning to turn red, as in my head I ran through all the times I had got it wrong. But before I had even finished apologising, he laughed and joked: "It's ok, I know we all look the same to you". Awkwardly I laughed along with him, knowing that any attempt at making excuses would just lead to further discomfort. My silence and awkwardness made him laugh, "I'm just fucking with you man", he reassured me. But from that moment on, my interactions with him changed. Until then, I had been taking it upon myself to walk up to him and ask him questions; always politely, but with no real concern about whether he *wanted* to speak to me, or what he thought of me. It was a stark reminder that I am not neutral or innocuous, especially in the eyes of Indigenous people. I am, unambiguously, a white man[13].

Ambiguously Indigenous

In the hoganesque structure of the Indigenous Pavilion, two men and a woman are seated behind a table. They are *Indigenous* because they wear feathered headdresses and beads around their necks and wrists, and they have patterns painted on their faces. The man in the middle is the most *Indigenous* because he is the oldest, wears the most makeup, and has the biggest feathers. He also does most of the talking. They are representatives of major IPOs from the Amazon and are discussing indigenous alternatives to REDD+. They talk about the threats faced by their peoples and the risks to their territories, asserting that *their* knowledge and *their* culture can protect *their* ancestral lands from destruction.

They are accompanied by another man who explains the importance of the project, REDD+ Indígena Amazónico (RIA), and how Indigenous

Peoples can protect the ecosystems that they live in by approaching conservation in a "holistic" way.

He goes into great detail about how and why we know that the presence of Indigenous Peoples helps to protect forests. His maps and measurements make sense to the small audience. He has a short animation: two maps that merge to show how patterns of deforestation go *around* indigenous territories in the Amazon. His evidence-based approach complements the impassioned and emotive speeches made by the Others on the panel, and the contrast is mutually legitimising. He is not Indigenous, or maybe he is. It is ambiguous because his hair is long, but it is grey, he is taller, and although not-quite-white, his skin is not brown enough to be sure. He wears no symbols or markers of indigeneity and is instead dressed in the standard "neutral" attire of this type of conference: a collared shirt, leather shoes, and smart trousers. If he is Indigenous, he is not being *Indigenous*. He is being professional: he seems like an "expert".

That it is sometimes unclear who is Indigenous and who is not is a reminder that being *Indigenous* in the Pavilion is dependent upon symbols and markers, and so is being an expert[14]. It is a place where Indigenous people who might otherwise be quite different from one another imagine, construct, and (re)present an apparently cohesive identity. This is done by Indigenous people in association with one another and requires certain aspects of Indigenous Peoples' cultures to be adapted to conform to Western symbolic notions of indigeneity. For the identity that emerges to be meaningful, it must necessarily exist in contrast to something: while *Indigenous* is primitive, passionate, emotional, and spiritual, not-indigenous (in this space, at least) is serious, professional, logical, and factual and is embodied and engendered by the symbolic markers of modernism and metropolitanism, including languages (particularly French, German, and Spanish, and especially English),[15] "professional" clothing and bodily presentation, and PowerPoint slides with maps and measurements. The environmental value of Indigenous Peoples and their territories is measured by pouring abstract units of tCO_2e into imaginary static landscapes where these so-called stewards live. It is reminiscent of how the intellectual capabilities of Indigenous people were once measured by pouring millet seeds into the skulls of dead Indians (Smith, 1999, p. 82), and this data and information are then presented by "experts", *Indigenous* or not, as if they are synonymous with understanding (Smith, 1999, p. 42) how and why this relationship between Indigenous people and the local environment has come to be.

One of the key reports (see Arrasco et al., 2014) that underpinned this presentation (and many others being made at the Indigenous Pavilion) was published by The Environmental Defense Fund (EDF) and The Woods

Hole Research Centre (WHRC). It was the product of a collaboration among IPO and NGO networks, scientists, and policy experts, including COICA[16]. The executive summary of *Tropical Forest Carbon in Indigenous Territories: A Global Analysis* stated that:

> The amount of CO_2 that would be released to the atmosphere if the forests in these territories were lost is 168.3 Gt CO_2 or more than 3 times the world's emissions in 2014. To continue to conserve the tropical forests that are essential to maintaining global climate stability, indigenous organizations need: 1. Titling of their territories as well as recognition of their rights to the vast natural resources of those territories and to the wealth of services they provide; 2. Relief from the persecution of their leaders who speak out in defense of indigenous rights and territories; 3. Recognition of the contributions of their people to climate change mitigation and adaptation and inclusion of those contributions in their governments' Intended Nationally Determined Contributions (INDCs); 4. Implementation of Free, Prior, and Informed Consent (FPIC) for forest conservation activities in indigenous territories; and 5. Direct access to climate financing for their organizations.
>
> *(EDF, 2015, p. 4)*

The report and the style of the presentation demonstrate two characteristics of this type of indigenous participation in environmental discussions: first, it suggests that in order for Indigenous Peoples to be considered "valuable" and for their (individual and collective) rights to be upheld, *it is necessary to demonstrate this value in terms of quantifiable properties that are of value to non-indigenous people* (tCO_2e, for example), and second that there is a tendency for Indigenous Peoples to be imagined in essentialist terms as a homogenous group with *shared characteristics that necessarily have a positive "contribution" to climate change mitigation*. Thus, for Indigenous Peoples to be treated as equal to non-indigenous people, they apparently must demonstrate their virtue and value as *Indigenous*: to be equal they must be *different*. In turn, this requires some Indigenous people to enter into partnerships with organisations like the EDF, The World Wildlife Fund (WWF), or The Nature Conservancy (TNC). The strategic value of working with "Big Green" and "pro-trade" environmental NGOs like these (Klein, 2014, p. 84) includes support from highly respected scientific research organisations (such as the WHRC), funding and technical assistance to develop "alternative" approaches, and the access necessary to disseminate these ideas, albeit in a highly translated format, in spaces like the Pavilion.

The Geodesic Dome

A year earlier, I had encountered the Indigenous Pavilion for the first time at COP20 in Lima, Peru. My first impression was that it was a sort of hybrid space that combined modern design with an idealised and romanticised vision of Indigenous Peoples' cultures. It was not obvious to me, though, whose vision of indigeneity it was, nor what purpose it served.

At the centre of a vast, open-sided hangar in the Voices for Climate Fair was a large, white, dome, made up of interlocking semi-opaque plastic triangles, the entrance to which was an arch surrounded by dense green foliage. On the side of the dome was a logo, three impressions of faces with exaggerated "Indian" features; white on a green backdrop, and the name of the exhibition space: "Pabellón Indígena". Among the flowers and leaves were two life-sized bronze statues – one seated, one standing – dressed only in loincloths and jewellery and accompanied by two large widescreen televisions. On the screens, images of Indigenous people smiling, canoes being paddled along rivers, exotic animals, and aerial images of landscapes played on a continuous loop.

Once inside the ultramodern Geodesic Dome, the dominant object was another life-sized exhibit: six more people, also cast in bronze, depicted a family or a tribe around a glowing fire. The woman in the sculpture was bare-breasted, reaching to the sky, which drew the eye towards the radiant Whiteness of the ceiling of the dome, and one of the men was snarling, pointing a spear down towards the fire. Around the edges of the space were more televisions showing similar images of forests and forest people, while sounds of nature (running water, insects and birds, and the occasional howling monkey) filled the air. A slow haunting song in a language I did not recognise, sung by a woman in a strained and emotive tone, occasionally cut through the natural soundscape, but it was unrecognisable enough to be only-just-human. The rest of the Dome was occupied by vertical white panels, supported by bamboo pillars, upon which there were more images of Indigenous people: smiling children in brightly coloured headdresses, close-ups of wrinkled brown hands weaving baskets, and women painting each other's arms in black ink. I had taken similar pictures myself when visiting a rainforest community and had posted them online, contributing in some small way to the collective curation of this shared notion of indigeneity, and people had commented that the portraits of people doing nothing unusual "looked like something from National Geographic",[17] which was meant as a compliment, I think.[18]

One particular image at the Pavilion stood out to me because it echoed the overall idea I felt was being conveyed in the space. It was taken from a high angle, looking down on two men, both shirtless and shoeless, wearing only shorts, seated at a tiny table playing chess while a young boy watched. The message communicated by this image, by the statues, and by the abundance of green and bamboo coupled with the crisp, white panels and the television screens, seemed clear: *two worlds are meeting here, and they are doing so harmoniously, for everyone's benefit.* "They" (Indigenous Peoples) can be and are becoming more like "us", and the "good" parts of their culture - the simplicity and the quaint exoticism, the spirituality and the connection to nature - can be conserved along with the environments in which they exist. But then there was the hyper-modern Geodesic Dome: the sharp, clean, and white, confines within which this encounter was taking place and by which it is restricted. It betrayed a subtler feature of the story being told in the Pavilion: that this apparently inclusive symbiotic relationship exists within a greater narrative of modernisation and development. This backdrop, though, is sanitised and whitewashed, and it carefully circumnavigates the violent and environmentally destructive history upon which the economies of the "so-called-developed-countries", to quote Manari, have been built.

Hybrid Identities in the Middle Ground?

The Indigenous Pavilions are ostensibly new and unique spaces where Indigenous Peoples encounter their Others, and for a short time, they are permitted (see Hale, 2004, pp. 16–17) to exist within a larger space that is shaped primarily by non-indigenous actors. On first observation, they appear to be places that have the potential to bring together ideas, ideologies, and perspectives that might otherwise be paradoxical: notions of Tradition and Progress, Conservation and Development, the Primitive and the Modern, the Past and the Future. And yet these places are not simply passive arenas created for the exchange of various knowledges, they are competitive spaces, a "marketplace of ideas"[19], the rules of which also exert an influence over which ideas can and cannot be discussed and how. Geographically, culturally, and aesthetically they are distant from most Indigenous Peoples, and yet for a short time each year a version of an Other world is imported. Here, a cultivated and calculated notion of what it means to be *Indigenous* at COP is portrayed to non-indigenous people.

This particular version of indigeneity cannot easily be described simply as a Western metropolitan fantasy (Tsing, 2007), nor is it an entirely "authentic" representation of the indigenous world. This is because, despite being a place that is, by definition, created by Indigenous Peoples, it exists within and for the purposes of the UN conference. The two Indigenous Pavilions I visited (Lima 2014 and Paris in 2015) were different to one another but shared some characteristics, and they are an interesting focal point from which to consider how and why Indigenous Peoples are (re)presented and (re)present themselves *as* the Other in non-indigenous spaces. The temporary existence of the Pavilions, as well as their physical position as an indigenous space nested within a non-indigenous event, mean that these transient "middle-grounds" become central to the process of (re)creating and (re)negotiating what it means to be *Indigenous* at COP, and in the globalised world more generally. The architecture, the discourses, the objects, and the bodily presentations exhibited become symbolic markers of this specific identity, constituting an emergent intersubjective indigeneity.

Places of this kind, where cross-cultural encounters between supposedly isolated societies can lead to "hybridity, syncretism, [or] cultural blending" (Davidann & Gilbert, 2013, p. 3), are sometimes referred to as "middle grounds", "hybrid spaces",[20] or "third spaces" that allow for cultural and epistemological disagreements to exist, and for negotiations between them to be somewhat meaningful (Torgerson, 2006, p. 725). As I noted, that was my first impression. But considering the Pavilion in these terms, particularly in terms of "hybridity", is not entirely sufficient and is somewhat problematic. So-called hybrid spaces are of interest because they are key to understanding modern societies and the interactions between them (Davidann & Gilbert, 2013, p. 2). But hybridity is a complex and contested term: it can be a "celebratory sign of diversity and mixedness", but it has also, at times, become a "monster of hegemony", being recruited in the discourse of neoliberal globalism and in the service of advancing global homogeneity (Bhabha, 2015, pp. x–xi). Moreover the idea of hybridisation brings to mind a process of mixing, as if the resulting phenomenon is an amalgamation of two or more existing things, and it euphemises the process by implying parity in the constituent parts of the whole. Similarly, "middle grounds" imply the possibility of a somewhat equitable encounter.

Notions of hybridity have historically also served to dehumanise colonial subjects, acting as the foundation of debates as to whether biological mixing – between racial groups that were at times seen as separate species – was possible at all. Later, assertions were made that, even if it is possible, hybridity would result in racial "decomposition" and diminished fertility (Young, 2005, pp. 8–13). This eventually informed sociological theories that feared social rather than physical degeneration. For example, Herbert Spencer, arguably the most renowned Western academic in the late 19th century, claimed that:

> Hybrid societies are imperfectly organisable – cannot grow into forms completely stable; while societies that have been evolved from mixtures

of nearly-allied varieties of man can assume stable structures, and have an advantageous modifiability.

(Spencer, 1876, cited in Young, 2005, pp. 17–18)

Thus, when considering people and places in terms of hybridity, it is necessary to pay attention to its history and its contradictions. Though it can be understood to "subvert and conflate long established classes and categories" (Werbner, 2015, p. 20), the Eurocentric and essentialist connotations of the term mean that the idea of hybrid social spaces evokes privileges and colonial aspiration, even when incorporating the more "desirable" elements of the culture of the Other (Papastergiadis, 2015, p. 261). At once, it can be seen as a means for political action and agency (see e.g., Hart, 2012, p. 146) and also as a scandal, intrinsically linked to notions of racial purity and apartness, and is therefore inherently transgressive (McClintock, cited in Goldberg, 2000, p. 72). Moreover, Papastergiadis (2015, p. 261) has observed that the conceptual transformation of hybridity from a dangerous sterilising process to a potentially productive tool does not mean it becomes benign. It may still serve, rather than challenge, the extension of the "European spirit" because it depends on a particular essentialist, modernist fantasy: assimilation through insemination – "the implanting of the white seed in the nurturing indigenous womb". Considering the Pavilion from both of these perspectives is useful, because it highlights the contradictory role of the space as inclusive and participatory, but simultaneously functioning to extend the "European spirit" and reduce the Other, (i.e., Indigenous Peoples) to the more desirable essentialised elements of their culture.

What such a space might mean in terms of a decolonised or postcolonial contemporary indigeneity is unclear. Frantz Fanon, the 20th-century Francophone Afro-Caribbean political philosopher, whose work has become influential in postcolonial studies, observed that "as" "it is the colonist who fabricated and continues to fabricate the colonized subject", decolonisation is not simply a preservation of the pre-colonial character of that subject, but instead fundamentally alters and transforms the colonised, bringing with it new ways of being, new language, and a new humanity that engenders the veritable creation of new people (Fanon, 1963/2004, p. 2). This is echoed by Silvia Cusicanqui who describes the project of indigenous modernity as having the potential to "emerge from the present in a spiral whose movement is a continuous feedback from the past to the future – a 'principle of hope' or 'anticipatory consciousness' – that both discerns and realises decolonization at the same time" (Cusicanqui, 2012, p. 96). Thus, in certain circumstances, this new and unique category of person, a decolonised, contemporary, Indigenous person who is neither "precolonial" nor a "hybrid", could come into being *because* of an antagonistic relationship with colonialism.

Ian Hacking's reflection on the ways in which categories of people are "made up" is pertinent here because it acknowledges that "the social reality is conditioned, stabilized or even created by the labels we apply to people, actions and communities" (Hacking, 1986, p. 226). Where the word "indigenous" is broad, taking on multiple meanings in varied contexts for wide variety of actors, *Indigenous* at COP is a far narrower and more specific label. And while it serves as a cohesive force among disparate people,[21] it simultaneously defines, and therefore limits, that which it is (and is not) to be *Indigenous*. Like the Canoe for Life, the people at the Pavilion are the embodiment of an intersubjective "true" Indigeneity; one that is a symbolic representation of an agreed upon category. Tsing's view on the "tribe" and becoming "tribal"[22] complements Hacking's observations on the role of categories. The argument put forward by Tsing – that categories such as these and the political actions that they engender lead to Indigenous Peoples reifying "Green Development Fantasies" and *becoming* them – runs alongside the point made by Hacking as it identifies that, in articulating knowledge about people within specific discursive categories, communities *become* identifiable objects to various outside actors (Tsing, 2007, p. 397): they *become* tribes. Yet Tsing exercises caution when critiquing the role of the category:

> The political rehabilitation of the tribe and its scholarly rejection too often speak past each other. [...] we must begin both our political rapproachments [sic] and scholarly investigations with how the concept of the tribe, with all its simplification and codifications of metropolitan fantasy, comes to mean something caught in particular political dilemmas.
>
> *(Tsing, 2007, p. 417)*

This type of political dilemma is all too visible at COP and particularly at the Pavilion. And the *Indigenous* people found there, along with the various symbols and markers of indigeneity, play no small part in assigning meaning to the category. Indigenous Peoples, if they wish to exert influence over negotiations pertaining to climate change, are implicitly required to demonstrate that they are, in fact, the ecological, spiritual, holistic, Indigenous Peoples of those Western, metropolitan, green fantasies: they must become *Indigenous*. The efficacy with which this is done is impressive, but it is also disquieting because, upon observation, it is always unclear which parts are "real" and which are "fantasy" (and indeed whose fantasies they are), and whether fantasy and reality are necessarily dichotomous. The question itself might be irrelevant. As Baudrillard noted, symbolic abstractions of this kind are not just the mirror of a concept but are instead models "of a 'real' without origin or reality: a hyperreal". His treatise on "Simulacra and Simulations" opens with a paraphrase of Ecclesiastes-

> The simulacrum is never that which conceals the truth – it is the truth which conceals that there is none.
>
> The simulacrum is true.

The "precession of simulacra" means (to take Baudrillard's example) that the map precedes the territory, and in the same sense, being *Indigenous* precedes being indigenous, in the context of COP at least, because:

> It is no longer a question of imitation, nor of reduplication, nor even of parody. It is rather a question of substituting signs of the real for the real itself; that is, an operation to deter every real process by its operational double, a metastable, programmatic, perfect descriptive machine which now provides all the signs of the real and short-circuits all of its vicissitudes.
>
> *(Baudrillard, 1981/2001, pp. 169–170)*

But Tsing's caution is important here, because it is a reminder that identifying this process need not undermine the political struggle of Indigenous Peoples by making a polemic scholarly rejection of their legitimacy at COP through accusations of inauthenticity, nor should it result in an uncritical political idolisation of Indigenous Peoples, which all too often resembles a kind of crypto-orientalism[23] or a faithful adherence to those "green development fantasies". Rather, this critique should draw attention to the *in*authenticity of non-indigenous efforts to "give voice" to Indigenous Peoples in climate change debates because such representation is always conditional. Moreover, it should serve as a reminder that Indigenous Peoples and indigeneity are not homogenous, and who speaks for them and what they say with regard to indigenous territories and the environment is also contested.

Conclusion

My initial observations at the Indigenous Pavilions in Lima and Paris provoked in me a degree of uneasiness and cynicism. The uncomfortable disposition I developed while conducting my research was, in part, brought on by observing the ways in which non-indigenous people tend to romanticise and/or dehumanise Indigenous people when talking about them, thinking about them, and interacting with them at COP and elsewhere. What made this particularly difficult to make sense of was that there appeared to be little effort on either side to challenge or even acknowledge this. Indeed, the *Indigenous* people themselves seemed to be encouraging it, albeit with an awareness of the pragmatic function of doing so. The key concern here is that, although it is transient and exists in this reified form *only* within the confines of the conference itself, the category of *Indigenous* (in line with the role of such categories, fantasies, or simulacra described by Hacking, 1986; Tsing, 2007; Baudrillard, 1981, respectively) is not benign. Though it has been argued that beneath the top-down essentialist classifications of races, identities, and "tribes" under colonialism there was (and still is) a subaltern reality defined in opposition to the imposed categories (Friedman, 2015, pp. 72–73), it is difficult to imagine that those who self-identify as Indigenous are entirely unchanged by the type of *Indigenous* identity made up at COP. Modern power, as understood by Foucault, need not be situated in central persons or institutions but is instead everywhere: it is "capillary", circulating throughout the entire social body to every extremity (Fraser, 1981, p. 272). From this perspective, though they are highly visible at the Indigenous Pavilions, the categorical schemes that relate to notions of indigeneity are produced and reproduced at all levels of society.

Though the Pavilion is described as an Indigenous Peoples' space, it conforms to characteristics that are specific to non-indigenous and decidedly "modern"[24] aesthetics: the Geodesic Dome, the arrays of televisions, and the structures of recycled wood and plastic. Or it recreates an imagined *Indigenous* world, complete with faux forests and images of Indigenous people in wild lands doing "primitive" things like hunting, fishing, and weaving. These are accompanied by displays of ritual acts and cultural, artefacts, such as ceramics, jewellery, musical instruments, and the Canoe for Life. This includes the people themselves who, when being *Indigenous*, embody many of the symbols associated with indigeneity, but when not doing so are practically indistinguishable from non-indigenous people. Werbner (2015, p. 20) describes this process, whereby a person changes through their self-presentation, stating that "Objects are the prosthetic extensions of a person; persons materially constituted by objects" and cites Latour's Actor Network Theory (ANT), which defines hybridity as the "product of mixed networks of people-objects". The two are indivisible, and yet some people-objects can be jarring. I noticed on a number of occasions non-indigenous people taking photos of *Indigenous* people who were wearing traditional clothing, but who were also using a mobile phone or a tablet. There is something compelling, or maybe entertaining, about this apparent cultural dissonance: it is a common but uncomfortable trope, as it carries with it a degree of Othering that verges on dehumanisation.

Seeing this happen – watching people taking pictures of people doing something completely mundane – brought to mind a photograph that I had seen at some point, which had won *Wildlife Photographer of the Year* in 2014. It showed a Japanese macaque using an iPhone. As the photographer noted in an interview, they "already resemble humans in so many ways, but when they're holding an iPhone, the similarities are almost scary" (Munoz, 2014). The juxtaposition of the primitive and modern, which echoes this idea – "they" are *almost* like "us", but not quite – is a thread that runs through the fabric of the Pavilion and seems to appeal to participants and observers alike. The result is that it becomes a space that simultaneously amplifies and mediates notions of sameness and Otherness, helping to both reproduce *and* redefine the category, as well as establishing ways in which Indigenous and non-indigenous identities can, and cannot, overlap. This occurs not as a singular moment, but as part of an ongoing historical and highly politicised process. It reaches back through time to when people who were described, and came to describe themselves, as Indigenous were colonised by European imperialist forces, and it cannot be easily separated from this narrative.

To imagine that indigenous perspectives and ideologies can interact with the hegemonic ideological framework in an entirely equitable, ahistorical manner in the process of making up the category of *Indigenous* in such a space would be naïve. As Escobar put it, "Power is too cyclical at the level of exploitation", and if the power of development discourse is to be challenged or displaced, the situation of exploitation must itself be recognised:

> The coherence of the effects that development discourse achieved is the key to its success as a hegemonic form of representation: the construction of the poor and underdeveloped as universal, preconstituted subjects, based on the privilege of the representers; the exercise of power over the Third World made possible by this discursive homogenization [...] and the colonization and domination of the natural and human ecologies and economies of the Third World.
>
> *(Escobar, 1995, p. 53)*

For this reason, imagining the Pavilion as a middle ground, or even as a hybrid space, is somewhat misleading. It can be better understood as a space where global hegemonic and (potentially) counter-hegemonic[25] forces encounter one another, a site of friction (see Tsing, 2011),[26] and though there is potential to subvert, and to assimilate in order to do so, this happens within the ideological framework of the coloniser. The power in such a place flows in both directions, but it is so asymmetrical that the mere presence of the more marginalised group, regardless of the restraints to which they must conform, is treated as a victory and is celebrated by Indigenous and non-indigenous actors alike. But this raises the question of why, if the power structure is so rigid, are Indigenous Peoples granted such a space at all? What, and who, are they there for?

Cusicanqui sheds light on the dynamic that shapes this moment in history, describing the present as "the setting for simultaneously modernizing and archaic impulses, of strategies to preserve the status quo [*aka pacha*] and others that signify revolt and renewal of the world: *Pachakuti*" but adds that discourses of hybridity and multiculturalism that risk supplanting Indigenous Peoples' struggles and demands, or that incorporate them into state apparatus in order to neutralise their will, represent "a change so that everything stays the same" (Cusicanqui, 2012, pp. 100–101). Thus, the presence of these *Indigenous* people (who are ostensibly the "leaders" of Indigenous Peoples) at COP functions in two ways: it *legitimises them* through formal recognition of their role as representatives of *all* Indigenous people, and *they in turn legitimise the negotiations that they enter into by virtue of being there at all.* This presence has the potential to either subvert or reinforce the status quo, or even do both: the role of being *Indigenous*, then, whether demonstrating openness or obstinance, might be, as Viveiros de Castro (2011, p. 73) put it, as inconstant as the Indian soul.

Notes

1 Of course, not all indigenous people have dark skin and dark hair, and wear makeup. But being *Indigenous* at COP appears to be associated with these characteristics. Some of the Northern European Sámi people, for example, have blonde hair, light skin and light eyes, and they wear less makeup (though some of the women paint thick read lines from the corners of their eyes to the tops of their ears). The Sámi do not look much like the indigenous people of the rainforest. Their forests are dry boreal, and their culture centres around herding, rather than hunting, but at COP they are identified as *Indigenous* and therefore within the collective identity of indigeneity, by other symbols. In particular, their clothing (the colourful felt Gákti and hats and boots lined with reindeer fur), their long hair, and their "primitive" musical instruments. Though they are "white", they are *Indigenous* because they are clearly *not* non-indigenous. As such, the Sámi occupy a particularly curious place within the category of *Indigenous*, in that they are potentially the least indigenous looking, and therefore could be viewed by some as less legitimate indigenous actors, but they are also white, and therefore could have a greater ability to transition between indigenous and non-indigenous identity, should they choose to do so.

2 Callison (2014) argues for establishing libraries in indigenous communities to act as a type of "new canoe", full of culture, language, art, identity, and knowledge that can be shared and passed through generations.
3 The canoe has both been traded and has facilitated the development of trade networks with and among Indigenous Peoples in the Americas. McSweeney (2004, p. 653) notes that the canoe "undermines the persistent notion that [...] engagement with markets [...] by indigenous people, is a fundamentally modern activity" (see also, Cusicanqui, 2012, p. 96, who notes that Indigenous Peoples were always, and are, contemporary beings and peers, not "pre-modern" societies).
4 The symbolic cultural significance of the Canoe is the foundation of a UNESCO CD-ROM aimed at Pacific youth with access to computers. Entitled *The Canoe is the People*, it provides an overview of canoe building, voyaging, and navigation and is "especially conceived to encourage Pacific youth to take pride in their heritage, and to keep their ancestral knowledge alive". It is described as "a critical step in revitalizing the transmission of indigenous knowledge". The education pack includes stories and documents that recount the cultural importance of the canoe in indigenous culture *to* indigenous people (UNESCO, 2014). An interesting metaphorical use of the canoe used by Michell (2012), a Cree researcher, is the 'Canoe Trip' which refers to doing community-based research, which he describes as "like going on a canoe trip to hunt for knowledge". Also see *The Metaphorical Transformations of the Wala Canoe* in Tilley (1999).
5 Goldman (1975, p. 64, cited in Viveiros de Castro, 1998, p. 133) puts canoes in the category "supernatural treasures" as part of the special sub-category of "containers" which also includes animal skins and houses (like the hogan), stating that the idea that all forms of life and vital force occupy a house or container is widespread among indigenous cultures throughout North and South America. The Kwakiutl, for example, speak of the body as the "house of the soul". Thus, the canoe qualifies as subject just as well as animals or manioc grinders (or even computers) do, having humanoid "embodiments" in the spirit world (Viveiros de Castro, 1998, p. 120).
6 A study conducted among the Maijuna people in Peru found that construction of one canoe required four 'mingas' (where members of the community come together to work on specific labour-intensive tasks). Each minga was 6 hours long and consisted of eight workers and, another 6 hours of individual labour was required to complete the canoe. During this process, at least 27 specific species (as well as others that were not specifically chosen for their species, but for a specific property) were used (Eshbaugh, Gilmore, & Greenberg, 2002, p. 20).
7 Reinharz called this the "rape model" of research (Reinharz, 1998, p. 95).
8 The Sarayaku produced a 34 minute film following COP21 "*Kawsak Sacha, La Canoa de La Vida*" (Kawsak Sacha, the Canoe of Life).
9 *Kava* is a drink made from the root of a pepper plant (*Piper methysticum*). It is a central part of various ritual ceremonies across Polynesia and it is connected to concepts of sacrifice. Its *mana* (power) is both pharmacological and cultural, and it is used to make contact with the spiritual world (Turner, 1986, p. 203).
10 For discussion on the role symbols (with reference to language, as opposed to cultural artefacts), Gillespie argues that "people spontaneously form intersubjective associations which are not simply intersubjective in the sense of being shared, but are intersubjective in the sense of weaving together divergent perspectives within the field of social action" (Gillespie, 2009, p. 34).
11 For an extensive discussion on the "Paradox of Authenticity", see Cobb, R. (2014). And for a more specific focus on how the concept of and paradoxical

nature of authenticity relate to structural colonial violence against Indigenous Peoples, see Maddison (2013).

12 The terms Zápara, Zaparos, and Sápara are used interchangeably in literature relating to the same indigenous group of Ecuador. For this reason, when quoting sources directly, I use the term chosen in the original text, but otherwise use "Sápara", as that is the term I have seen and heard the people themselves use most frequently.

13 I use the term white man here to intentionally make explicit my position as Other from the perspective of the people I am interacting with. Though it may appear to be obvious or trivial, it is not. The act of categorising indigenous people *requires* that I, as a non-indigenous person, be categorised, too. Most often, the term I have heard indigenous people use to categorise non-indigenous people (in English, at least) is "white man" and so I use it hear for that reason. This is a small step toward treating whiteness not as neutral, but as a crucial part of interactions, and doing the "critical and self-reflexive ideological work necessary to make whiteness visible and overturn its silences for the purpose of resisting racism" in order to make explicit "unspoken issues of race, [in] discursive spaces where the power of whiteness is invoked but its explicit terminology is not" (Crenshaw, 1997, p. 254).

14 Kothari notes that so-called experts "embody the unequal relationship between the 'First' and 'Third' Worlds, and between donors and aid recipients, and exemplify the process through which development is located within institutionalised practices. This production of the 'professional' development expert, identified as such not solely because of the extent and form of their knowledge but often because of who they are and where they come from, legitimises and authorises their interventions by valorising their particular technical skills and reinforcing classifications of difference" (Kothari, 2005, p. 426).

15 As Mignolo notes, Spanish ranks lower in the epistemological hierarchy, meaning that even when speaking the language of the colonizer, Latin American indigenous people face yet another degree of marginalisation "Spanish language is today, in the European Union, less influential and less sustainable than knowledge produced in English, French or German – English, above all due to the imperial leading role of the US" (Mignolo, 2009, p. 73).

16 See following chapter.

17 National Geographic has, in recent years, begun to reflect on its own role in creating and perpetuating racist stereotypes and has attempted to shift from presenting "the 'natives' elsewhere as exotics, famously and frequently unclothed, happy hunters, noble savages – every type of cliché". And acknowledged that the publication had done "little to push its readers beyond the stereotypes ingrained in white American culture" (Goldberg, 2018). Interestingly, the last time such an image was presented on the cover of the magazine was on the January 2014 issue. The cover was a close-up image of a young man, with long black hair, brown skin, beads around his neck and black, and red geometric lines painted on his face, accompanied by the headline "Defenders of the Amazon: Taking on the Modern World and Winning" (National Geographic, 2013). Some of the Kayapo people pictured in the article inside the magazine were also in attendance at COP20 and COP21.

18 I did not take it as a compliment. For a long time, I have found *National Geographic* an uncomfortable read. Its tendency toward essentialism, exoticism, and Othering is off-putting to say the least. As Parameswaran argues, its discursive strategies have long been entangled with "hierarchical structures of gender, race, and class [...and a...] neocolonial rendering of global culture" (Parameswaran, 2002, p. 288).

19 I use the term "marketplace of ideas" here intentionally, with an awareness of its problematic connotations. It assumes the best ideas will be most successful and is intrinsically linked to neo-liberal democracy and market fundamentalism. But although it is based on "European values, practices and structures that often do not fit different cultural and political histories of the postcolony" (Chiumbu, 2013, p. 71) it is the correct analogy, for precisely that reason, in the context of COP.

20 "Middle ground" and "hybrid space" are often used interchangeably, and both are linked to and similar in meaning to Bhabha's term "third space" (1990) and are at times used interchangeably in various fields (see e.g., Kulle, 2003. pp. 95–120; López-Robertson & Schramm-Pate, 2013, pp. 40–56; Voicu, 2014, pp. 125–132).

21 Though the implicit (or explicit) goal of a relatively homogeneous pan-Indigenous social and political community may be problematic, it has been suggested that this haven of pan-Indigeneity is a prerequisite for indigenous people to "resist the seduction of assimilation and confidently work at rebuilding a unique identity" (Ariss, 1988, cited in Paradies, 2006, p. 356).

22 "Tribe" and "indigenous" are similar in meaning and are often treated as synonyms, though the latter has largely replaced the former in international political discourse. Along with "native", "aborigine" "first nations" and "Indians", the uses of the terms have come in and out of favour in different places and times. For discussion on the various connotations of these terms, see Mika and Peters (2017).

23 All scholars, according to Gomez, "must heed the danger signs of crypto-Orientalism – the willingness to bask in the glory of our texts and then use them to our own ends, the desire to tell our subjects what they really think, and the compulsion to deny any sympathetic involvement" (Gomez, 1995, p. 229).

24 Cusicanqui (2012, p. 107) rejects the notion that being indigenous and being modern are dichotomous, arguing that Indian hegemony can be "realized in spaces that were created by the cultural invader: the market, the state, the union. In doing so, we create our own project of modernity, a more organic one than that imposed by the elites". Conversely, Latour, made a similar suggestion that the "doubly asymmetrical" characteristics ascribed to modernity mean that "we moderns" "have never been modern" (Latour, 1993, pp. 10 and 46).

25 For reflections on Gramscian hegemony and counter hegemony, see e.g., Carroll and Ratner (1994) and, for a more critical reading of the misuse of counter hegemony, in the development world see McSweeney (2004). A more radical, decolonial, reading of the concept hegemony and the possibility of an "indigenous hegemony" that confronts "the hegemonic projects of the North" and that can potentially be realised in the spaces "that were created by the cultural invader" can be found in Cusicanqui (2012).

26 Tsing describes global connections through the metaphor of "friction" (which is not a synonym for resistance) that demonstrates the grip of the encounter: "A wheel turns because of its encounter with the surface of the road; spinning in the air it goes nowhere. Rubbing two sticks together produces heat and light; one stick alone is just a stick. As a metaphorical image, friction reminds us that heterogeneous and unequal encounters can lead to new arrangements of culture and power" (Tsing, 2011, p. 5). It is crucial to acknowledge though that friction destroys just as well as it creates and that knowing who is the wheel and who is the road gives some indication of what the outcome of such encounters might be.

References

Arrasco, C. S., Baccini, A., Augusto, C., Jintiach, J. C., Meyer, C., Miranda, M. A., Ricardo, B., Ríos, S., Ruiz, M. R., Schwartzman, S., Smith, R., Soria, C., Vasquez, E., & Walker, W. (2014). Forest carbon in Amazonia: The unrecognized contribution of indigenous territories and protected natural areas. *Carbon Management*, *5*(5–6), 479–485.

Baker, L., Brock, S., Dove, M. R., Hebdon, C., & Ludlow, F. (2016). The double binds of indigeneity and indigenous resistance. *Humanities*, *5*(3), 53.

Baudrillard, J. (1981/2004). Simulacra and simulations. In M. Poster (Ed.), *Selected writings* (pp. 169–187). Cambridge: Polity (original work published 1981).

Bhabha, H. K. (2015). Foreword. In P. Werbner & T. Modood (Eds.), *Debating cultural hybridity: Multicultural identities and the politics of anti-racism* (pp. ix–xiii). London: Zed Books.

Callaway, S. M. (1968). *Grandfather Stories of the Navahos*. The Board of Education: Rough Rock Demonstration School. Retrieved from https://files.eric.ed.gov/fulltext/ED037289.pdf

Callison, C. (2014). Indigenous peoples' new canoe. In D. Lee, & M. Kumaran (Eds.), *Aboriginal and visible minority librarians: Oral Histories from Canada* (pp. 135–146). Plymouth: Rowman & Littlefield.

Carlson, B., Harris, M., & Poata-Smith, E. S. (2013). Indigenous identities and the politics of authenticity. In B. Carlson, M. Harris, & M. Nakata (Eds.), *The politics of identity: Emerging indigeneity* (pp. 1–9). Sydney: University of Guadalajara Press.

Carroll, W. K., & Ratner, R. S. (1994). Between Leninism and radical pluralism: Gramscian reflections on counter-hegemony and the new social movements. *Critical Sociology*, *20*(2), 3–26.

Cepek, M. (2012). *A future for amazonia: Randy borman and cofán environmental politics*. Austin: University of Texas Press.

Chiumbu, S. (2013). Knowledge and Ideas in a Context of Power: Rethinking media policy and reform in southern Africa. *Africanus*, 43(2), 64–77.

Cobb, R. (Ed.). (2014). *The paradox of authenticity in a globalized world*. New York: Palgrave.

Conklin, B. A. (1997). Body paint, feathers, and VCRs: Aesthetics and authenticity in amazonian activism. *American Ethnologist*, *24*(4), 711–737.

Conklin, B. A., & Graham, L. R. (1995). The shifting middle ground: Amazonian Indians and eco-politics. *American Anthropologist*, *97*(4), 695–710.

Crenshaw, C. (1997). Resisting whiteness' rhetorical silence. *Western Journal of Communication*, *61*(3), 253–278.

Cusicanqui, S. R. (2012). Ch'ixinakax utxiwa: A reflection on the practices and discourses of decolonization. *South Atlantic Quarterly*, *111*(1), 95–109.

Davidann, J. T., & Gilbert, M. J. (2013). *Cross cultural encounters in modern world history*. New York: Routledge.

Denzin, N. K., Lincoln, Y. S., & Smith, L. T. (Eds.) (2008). *Handbook of critical and indigenous methodologies*. London: Sage.

Duhalde, C., & Vallejo, I. (2017). Políticas de desarrollo y ampliación de las fronteras extractivas sobre territorio Sapara en la Amazonía Ecuatoriana. *Abya-yala: Revista sobre Acesso à Justiça e Direitos nas Américas*, *1*(1).

Duranti, A. (2010). Husserl, intersubjectivity and anthropology. *Anthropological Theory*, *10*(1-2), 16–35.

EDF (2015). *Tropical Forest carbon in Indigenous Territories: A Global Analysis*. Retrieved from https://www.edf.org/sites/default/files/tropical-forest-carbon-in-indigenous-territories-a-global-analysis.pdf

Escobar, A. (1995). *Encountering development: The making and unmaking of the third world*. Princeton: Princeton University Press.

Eshbaugh, W. H., Gilmore, M. P., & Greenberg, A. M. (2002). The use, construction, and importance of canoes among the Maijuna of the Peruvian Amazon. *Economic Botany*, *56*(1), 10–26.

Fanon, F. (2004). *The wretched of the earth* (R. Philcox, Trans.). New York: Grove Press. (original work published 1963).

Fraser, N. (1981). Foucault on modern power: Empirical insights and normative confusions. *Praxis International*, *1*(3), 272–287.

Friedman, J. (2015). Global crises, the struggle for cultural identity and intellectual Porkbarrelling: Cosmopolitans versus locals, ethnics and national in an era of de-homogenisation. In P. Werbner, & T. Modood (Eds.), *Debating cultural hybridity: Multicultural identities and the politics of anti-racism* (pp. 70–89). London: Zed Books.

Gillespie, A. (2009). The intersubjective nature of symbols. In B. Wagoner (Ed.), (2010), *Symbolic transformation: The mind in movement through culture and society* (pp. 23–37). London: Routledge.

Goldberg, D. T. (2000). Heterogeneity and hybridity: Colonial legacy, postcolonial heresy. In H. Schwartz, & R. Sangeeta (Eds.), (2005). *A companion to postcolonial studies* (pp. 72–86). London: Blackwell.

Goldberg, S. (2018). For Decades, Our Coverage Was Racist. To Rise Above Our Past, We Must Acknowledge It. *National Geographic: The Race Issue* [Online]. Retrieved from https://www.nationalgeographic.com/magazine/2018/04/from-the-editor-race-racism-history/

Goldman, I. (1975). The mouth of heaven: an introduction to Kwakiutl religious thought. New York: Wiley Interscience.

Gomez, L. O. (1995). Oriental wisdom and the cure of souls: Jung and the Indian East. In D. S. Lopez (Ed.), *Curators of the Buddha: The study of Buddhism under colonialism* (pp. 197–250). London: University of Chicago Press.

Hacking, I. (1986). Making up people. In D. E. Wellbery, M. Sosna, & T. C. Heller (Eds.), *Reconstructing individualism* (pp. 222–236). Stanford: Stanford University Press.

Hale, C. (2004). Rethinking indigenous politics in the era of the "Indio Permitido". *NACLA Report on the Americas*, *38*(2), 16–21.

Hart, J. (2012). Hybridity and being between cultures. *Canadian Review of Comparative Literature/Revue Canadienne de Littérature Comparée*, *39*(2), 139–151.

Hartigan, J. (2013). Anthropology of race. In J. Hartigan (Ed.),, (Ed.). (2013). *Anthropology of race: Genes, biology, and culture*. Santa Fe: SAR Press.

Jacome, F. (2015, January 24). Amazon Guardians of Life: Fighting Against Big Oil. *Third Eye*. Retrieved from http://www.thethird-eye.co.uk/guardians-of-life-amazon-women-ecuador-by-felipe-jacome/

Klein, N. (2014). *This changes everything*. London: Random House.

Kothari, U. (2005). Authority and expertise: The professionalisation of international development and the ordering of dissent. *Antipode*, *37*(3), 425–446.
Kulle, D. (2003). Where's the middle? The 'Less-Aligned' in a context of absolutes: An anthropological perspective on identities in Northern Ireland. *Nordic Irish Studies*, *2*, 95–120.
Lane, L. (1999). Practical and religious meanings of the navajo hogan. In V. D. Nazarea (Ed.), *Ethnoecology: Situated knowledge/located lives* (pp. 37–45). Tucson: University of Arizona Press.
Latour, B. (1993). *We have never been modern*. Cambridge: Harvard University Press.
Li, T. M. (2000). Locating indigenous environmental knowledge in Indonesia. In A. Bicker, P. Parkes, & R. Ellen (Eds.), *Indigenous environmental knowledge and its transformations: Critical anthropological perspectives* (pp. 121–150). London: Harwood.
López-Robertson, J., & Schramm-Pate, S. (2013). (Un) official knowledge and identity: An emerging Bilingual's journey into hybridity. *Innovation in Language Learning and Teaching*, *7*(1), 40–56.
Maddison, S. (2013). Indigenous identity, 'authenticity' and the structural violence of settler colonialism. *Identities*, *20*(3), 288–303.
Manuelito, K. (2005). The role of education in American Indian self-determination: Lessons from the Ramah Navajo community school. *Anthropology & Education Quarterly*, *36*(1), 73–87.
McSweeney, K. (2004). The dugout canoe trade in Central America's Mosquitia: Approaching rural livelihoods through systems of exchange. *Annals of the Association of American Geographers*, *94*(3), 638–661.
Michell, H. (2012). The Canoe Trip: A Northern Cree Metaphor for Conducting Research. Retrieved from https://journals.uregina.ca/ineducation/article/view/4
Mignolo, W. (2009). Dispensable and bare lives: Coloniality and the hidden political/economic agenda of modernity. *Human Architecture*, *7*(2), 69.
Mika, C. T., & Peters, M. A. (2017). Aborigine, Indian, indigenous or first nations? *Educational Philosophy and Theory*, *49*(13), 1229–1234.
Munoz, G. (2014). *Gallery: Celebrating 50 years of Wildlife Photographer of the Year*. Retrieved from https://www.sciencealert.com/gallery-celebrating-50-years-of-wildlife-photographer-of-the-year
Naku (n.d.). *Even if we disappear*.
NakuNorth (n.d.). *Journeys: Itinerary*. Retrieved from http://nakunorth.com/journey-2018/
National Geographic (2013, December 6). January 2014 Edition, 'Defenders of the Amazon: Taking on the World and Winning'. *National Geographic*. Retrieved from http://press.nationalgeographic.com/2013/12/16/national-geographic-magazine-january-2014/
Papastergiadis, N. (2015). Tracking hybridity in theory. In P. Werbner, & T. Modood (Eds.), (2015). *Debating cultural hybridity: Multicultural identities and the politics of anti-racism* (pp. 257–282). London: Zed Books.
Paradies, Y. C. (2006). Beyond black and white: Essentialism, hybridity and indigeneity. *Journal of Sociology*, *42*(4), 355–367.
Parameswaran, R. (2002). Local culture in global media: Excavating colonial and material discourses in national geographic. *Communication Theory*, *12*(3), 287–315.

Pedersen, M. A. (2011). *Not quite shamans: Spirit worlds and political lives in Northern Mongolia*. London: Cornell University Press.
Pedersen, M. A., & Willerslev, R. (2012). "The soul of the soul is the body" re-thinking the concept of soul through North Asian ethnography. *Common Knowledge*, *18*(3), 464–486.
Praet, I. (2005). People into ghosts: Chachi death rituals as shape-shifting. *Tipití: Journal of the Society for the Anthropology of Lowland South America*, *3*(2), 3.
Praet, I. (2009). Shamanism and ritual in South America: An inquiry into Amerindian shape-shifting. *Journal of the Royal Anthropological Institute*, *15*(4), 737–754.
Reinharz, S. (1998). *On becoming a social scientist*. London: Transaction Publishers.
Salazar, N. B. (2013). Imagineering otherness: Anthropological legacies in contemporary tourism. *Anthropological Quarterly*, *86*(3), 669–696.
Smith, L. T. (1999). *Decolonizing methodologies: Research and indigenous peoples*. London: Zed Books.
Taussig, M. T. (1987). *Shamanism, colonialism, and the wild man: A study in terror and healing*. London: University of Chicago Press.
Tilley, C. (1999). The metaphorical transformations of Wala Canoes. In V. Buchli (Ed.), *The material culture reader* (pp. 27–55). London: Routledge.
Torgerson, D. (2006). Expanding the green public sphere: Post-colonial connections. *Environmental Politics*, *15*(5), 713–730.
Tsing, A. L. (2007). Becoming a tribal elder, and other fantasies of green development (reprint). In M. Dove, & C. Carpenter (Eds.), *Environmental anthropology: A historical reader*. Malden, MA: Blackwell.
Tsing, A. L. (2011). *Friction: An ethnography of global connection*. Woodstock: Princeton University Press.
Turner, J. W. (1986). The water of life: Kava ritual and the logic of sacrifice. *Ethnology*, *25*(3), 203–214.
UNESCO (2014). *The Canoe Is the People: UNESCO Interactive Educational Resource*. Retrieved from http://un.org.au/2014/03/17/the-canoe-is-the-people-unesco-interactive-educational-resource/
Viatori, M. (2005). New bodies, ancient blood: "Purity" and the construction of Zápara identity in the Ecuadorian Amazon. *Journal of the Society for the Anthropology of Lowland South America*, *3*(2), 175–195.
Viveiros de Castro, E. B. (1998). Cosmological perspectivism in Amazonia and elsewhere. *HAU Journal of Ethnographic Theory, Masterclass Series*, *1*, 45–168.
Viveiros de Castro, E. B. (2011). *The inconstancy of The Indian soul: The encounter of Catholics and cannibals in 16th century Brazil*. Chicago: Prickly Paradigm Press.
Voicu, C. G. (2014). *Exploring cultural identities in Jean Rhys' fiction*. Berlin: De Gruyter.
Werbner, P. (2015). Introduction: The dialectics of cultural hybridity. In P. Werbner, & T. Modood (Eds.), *Debating cultural hybridity: Multicultural identities and the politics of anti-racism* (pp. 1–28). London: Zed Books.
Young, R. J. (2005). *Colonial desire: Hybridity in theory, culture and race*. London: Routledge.

4

ENCOUNTERING THE OPTIMAL OTHER

Introduction

In the Andean mountains, 2,850 metres above sea level, Quito stretches to within a kilometre of the Equator. It is still in the southern hemisphere but only just. The capital city of Ecuador is, in many ways, a boundary between worlds. It serves as a frontier upon which the ideologies and interests of various actors from Ecuador and from elsewhere converge. Sometimes these are combined or exist in parallel, but at other times they come into conflict. Historically, the more dominant ideology has at times violently suppressed that of the less powerful groups. The most notable example of this, as with most of Latin America, is the dominance of the Spanish conquistadors over the "pre-Columbian" people of the region. This is still visible today in the colonial architecture of the city, the numerous Catholic Churches, and the streets and squares that bear the names of the greatest and most revered of the colonisers.

For the majority of my time in Ecuador, I lived on a street called *Camino de Orellana*, overlooking the *Avenida de los Conquistadores* (Avenue of the Conquerors), which was built along the route that Francisco de Orellana travelled from Quito down into the Amazon rainforest, where he completed the first recorded navigation of the entire Amazon River.[1] At the bottom of the hill stands a statue of Orellana in a suit of armour, sword in hand, looking to wards the seventeenth-century Catholic Church, *El Santurio de Guápulo*, which, I was told by my neighbours, was built using the labour of Indigenous and Afro-descendent enslaved people.

DOI: 10.4324/9781003341864-7

The plaque reads "From the Government of the Republic [...] to Francisco de Orellana. Discoverer of the Amazon. 1995".

Although the ways that it happens have changed over time, Quito remains a battleground upon which various groups seek to resolve conflicts.[2] The geography and architecture of the city serve as reminders that some actors and ideas begin these battles from a far more powerful position than others. In the *Plaza Grande* of the Old Town, the presidential palace - an expansive, white, colonial building - today remains a site of protests, particularly for Indigenous Peoples, that at times have resulted in violent clashes between protestors and police or armed forces.[3] But it is against this backdrop that some Indigenous Amazonian people are carving out a space for themselves in the political world in order to have a say in their Peoples' future and the ways the forest in which they live will be managed.

Near my home in Guápulo was a quiet and pretty part of Quito called La Floresta, where, on of January 7, 2015, I arrived for the first time at the head office of Coordinator of Indigenous Organizations of the Amazon River Basin (COICA). It is among the most important Amazon Indigenous Peoples' Organisations (IPOs) at the international level and takes a leading role at the Indigenous Pavilion at COP. It is the only coordinator of the various indigenous federations from the nine countries whose territories reach into the Amazon basin, but at the national level it is far less visible to the general public than the organisations it represents.[4] COICA exists primarily in the international arena, with representatives attending important events around the world, interacting with governments, the UN, the World Bank, and numerous major environmental Non-Governmental Organisations (NGOs), and it attracts significant funds from donor organisations. In this sense, it is quite unusual, and its unique character as a bridge between indigenous and non-indigenous worlds is reflected in its location.

The quiet, unassuming, office in a picturesque neighbourhood of a metropolitan city makes it easy to forget that the COICA ostensibly represents hundreds of thousands of Amazonian people, including those in voluntary isolation. But this is, nonetheless, what it does. Up here in the mountains, on the very edge of the geographical and political Global North, looking down over the *Avenida de los Conquistadores*, COICA occupies the highest political level of the Amazon Indigenous Peoples' struggle, and yet it is both physically and culturally distant from the places and people it represents. In order to do the work that it does it has to be. This location also evokes a key idea that is discussed throughout this book: because it is here that this relatively small number of Indigenous people take on the task of representing *all* Indigenous Peoples: a highly diverse set of communities, nationalities, and ethnicities. They also seek to integrate concepts that are central to indigenous cosmovisions into the framework of the dominant or mainstream approaches to "sustainable development".

This is no modest undertaking because to become part of this world where the most powerful actors discuss and negotiate the future of the Amazon rainforest, leaders and representatives of Indigenous Peoples must become separate from their communities, homes, and families and navigate and negotiate the world of international development. This in turn raises concerns of legitimacy and potentially opens COICA up to questions relating to their "authenticity". In conversations with national and international NGOs that also purportedly represent the interests of Ecuadorian Indigenous Peoples in various capacities, this concern was raised to me in the following way by a white man from the United States who worked for a US NGO:

> On the one hand, given the power of governments or companies, to have an entity that can lend its voice to pan-Amazonian indigenous issues, there's strength in that, but at the same time there's danger, and we see it playing out right now. When you have interests from governments or NGOs, or even, potentially, companies, there's a danger in the idea that somehow COICA could speak for or represent all Amazonian people. I think it's super problematic.
>
> *(D., 2015)*

How can a small number of people who are so far from those they represent speak and act on their behalf? It is a fair question, but paradoxically it is precisely this location and the political position of the organisation that allows COICA to connect indigenous Amazonian federations to one another and, subsequently, to numerous influential organisations from the Global North. It also presents the possibility of Indigenous Peoples having a physical presence and a collective space at events like the United Nations Framework Convention on Climate Change (UNFCCC)-COP (see Chapter 3).

The Strategy of Being "Good" Subjects

The phenomenon of some Indigenous or colonised people existing in a state of Otherness from their community is not unique to COICA. As Dion-Buffalo and Mohawk, note, there are limited options available:

> They can become "good subjects" of the discourse, accepting the rules of law and morals without much question, they can be "bad subjects" arguing that they have been subjected to alien rules but always revolting within the precepts of those rules, or they can be "non-subjects" acting and thinking around discourses far removed from and unintelligible to the West.
>
> *(Dion-Buffalo & Mohawk, 1992, p. 19)*

Arguments for and against taking any of these positions can (and have been[5]) be made. But only the first of the three allows for any possibility of direct agency within the political institutions of the Global North, at least in the short term. And so, one could equally ask who, if not an organisation like COICA, could ensure that Indigenous Peoples of the Amazon have a voice in these types of forums? If nothing else, COICA as an organisation appears to be highly skilled at speaking in the language of the dangerous Other world, identifying with all that is condemned, and appearing, at least, to be "good subjects".[6] This shift in identity, Othering oneself for the benefit of one's community, is not a uniquely indigenous problem; it is encountered by many people from subaltern communities who become part of mainstream political processes. Mookherjee (2010), for example, describes this predicament whereby to facilitate the struggle to transcend the politically subordinate position, the subaltern has two potentially problematic courses of action. One is the strategic struggle for group recognition *within* the institutional context, which risks both confirming their subordinate position within it and strengthening essentialist notions *about* the group. The other is based on subversion and resistance, but as Mookherjee notes, it is unclear how struggle of this kind "could be socially transformative without resorting to revolution or violent protest" (Mookherjee, 2010, p. 185).

This chapter discusses the ways in which indigeneity appeared to me among the people with whom I interacted during my time with COICA. Of particular interest is the ways in which it is and is not evoked at different times and how it is employed to assert agency and influence in "mainstream" discussions that focus on the forests in which they, or their families and communities, and the many more people they represent, often live. Central to this discussion is the position of the organisation on a frontier between the indigenous worlds of the Amazon and the world of sustainable global development, which is founded on discourses that radiate outward from the development organisations, governments, and intergovernmental organisations of the Global North. I build upon the observations of the previous chapter to illustrate the ways in which notions of being *Indigenous* (that are intimately connected to such discourses) can flow into and through, and are (re)shaped by, Indigenous Peoples.[7] Its emergence as a significant actor in debates around forests and climate change are a testament to the pragmatism of COICA, but it appears to have affected the organisation and the people it represents in two ways. First, they have become *less* like the *Indigenous* people imagined by non-indigenous people, changing the ways that they (re)present themselves and interact with their Others (including me) in order to fit within the epistemological framework of "mainstream" approaches by becoming apparently more "professional" and "rational"; and second, they become *more* like the Indians of "green development fantasies" (Tsing, 2007) by strengthening their commitment to "ecological", "sustainable", and "holistic" approaches to forests and emphasising their role as "stewards". They do so with the aim of securing territorial rights, which is presented by COICA as a fundamental part of ensuring their cultural (and physical) survival. From the outside, they appear to become less "savage" and more "ecologically noble".

*

From the moment I arrived at COICA, I was granted open access to the office. I set up my desk the Monday following my first visit, as we had agreed, and from then on, I came and went as I pleased. I spent much of the following eight months in this often-empty office, and though I had more access to people, events, meetings, and documents than I had expected, it was such a "normal" and mundane space that I rarely felt as though I was doing research at all.[8] Getting interviews with people was difficult – always "*mañana, mañana*" – and the *pasantía* Tomás had suggested never materialised. Though I wasn't ignored, as such, I felt like my presence in the office went largely unnoticed by most of the people who came and went. But as with the exam described int Chapter 1, my English language and translation skills, as well as my supposed position as an "expert", would occasionally be called upon. Requests ranged from writing, editing, or translating of documents and web pages (many of which were never used for anything), taking photos of COICA staff for visas and assisting with application processes, and sending emails or making phone calls, including a request to contact Prince Charles and invite him to visit COICA (which I thought at the time was a joke at my expense but I tried to do it anyway). I also tried to make myself useful by bringing in coffee and *bolones* (plantain dumplings) in the morning or by helping move furniture around for meetings and events. Most of the staff were friendly, all were polite, and soon they were making jokes about the British and gringos, and occasionally asking me questions about why I was there. I felt comfortable and safe in the space.

In many ways, it was just like any other head office of a business or NGO that might be found in Quito, or any other city in the world for that matter. The staff was a mix of Indigenous and non-indigenous people, and the building was utilitarian: concrete and steel, wooden floors, single-paned windows with beige vertical blinds, and white walls. There appeared to be little that marked the office out as an indigenous space, except for a row of portraits of rainforest people that hung along the wall in the conference room, and two maps displayed alongside one another near my desk in the communal working space. The first was a standard political map of South America, but the second was a carbon map of the Amazon Rainforest. It was produced by COICA in collaboration with Woods Hole Research Center (WHRC), Environmental Defense Fund (EDF), and the Amazonian Network of Geo-referenced Socio-Environmental Information (RAISG)[9] to visualise the inverse relationship between Indigenous Peoples' territories and deforestation, and the impact this has on the carbon that is stored in the Amazon. The map is used by COICA to demonstrate that patterns of deforestation go *around* indigenous territories, and that they are even more effective at halting deforestation than Natural Protected Area (NPAs) (see RAISG, 2012).

The people, though, brought a character to the office that felt, to me, distinctly indigenous. Though most communication in the office was in Spanish, a variety of indigenous languages were spoken at different times. "*Pachakuti*"[10] one member of staff would shout almost every time he entered the office, and now and then *cumbia* music would flow through the building from the keyboard in Tomás's office. Occasionally, the building would act as a stopping point for people travelling from the rainforest to the city for meetings or other administrative reasons, or for medical care. They would sit for long periods of time and, occasionally, people would sleep there, sometimes in groups, sometimes alone. I spoke with the people who passed through sometimes, though some did not seem at ease with me or did not speak Spanish, and I was careful to not make them uncomfortable by forcing conversation. Sometimes, though, larger groups would turn up before and during protests that were being held in the capital. Although COICA usually tried to accommodate them, when protests came to a head during the indigenous uprising in 2015, the office was shut down for a week to avoid becoming entangled in the conflict that was emerging between other indigenous organisations (particularly the Confederation of Indigenous Nationalities of Ecuador [CONAIE])[11] and the Government of Ecuador.

The office of COICA was a place with multiple functions. It acted as a meeting point and a safe space for Amazonian Indigenous Peoples in the city, but it was also constrained in this role by the need to remain politically neutral and maintain its character as a bridge between worlds. From day to day, indigeneity was not deployed in the office in the way that it is described Chapter 4: it did not seem necessary for the people to demonstrate their adherence to the symbolic representations of what it means to be *Indigenous*. The character of indigeneity and its apparent role was different than at the Indigenous Pavilion. Where at the Pavilion it is savvy to be more *Indigenous*, in this place, it made sense to be more *Professional*. The space is permanent and continuously populated by people who are themselves Indigenous and by Others who (like me) have to some extent already recognised COICA's legitimacy as representatives of Indigenous Peoples and have chosen to interact with them in their space, implicitly acknowledging their indigeneity, however it is presented, and the indigeneity of the space itself. In their own space, members of COICA work with non-indigenous people in a professional, as opposed to symbolic, capacity, which includes managing significant funds, implementing projects, and coordinating at the international level – thus, demonstrating a capacity to fulfil these requirements through *sameness* is more pragmatic than demonstrating *difference*.

One of my first thoughts when I began spending time with COICA was that they were indeed very capable and very professional (surprisingly so, based on my previous assumptions and limited experiences with Indigenous Peoples), which I implicitly equated to being *less* "authentically" indigenous. Though COICA being indigenous is why we Others come, being "professional" is why we stay. The need to communicate or translate ideas and concepts to non-indigenous people in ways that can be understood, and that are deemed "appropriate" and "effective", is actively being learned and taught within the organisation. And yet, at times, particular aspects of indigeneity, including bodily presentations, rhetoric, and performances, some of which appear "authentic" and others that do not, are deployed, particularly when representatives of COICA are interacting with one another and with other Indigenous people.

This chapter is concerned with how and when this does and does not happen, and what it means for the role of COICA as "leaders" or "representatives" of Indigenous Peoples and as agents in the construction of indigeneity. I consider this through my observations of three events that I attended while spending time with COICA. I begin with COICA's Anniversary Conference, which brought together members of the organisation to celebrate 30 years since its inception in 1984. I then reflect on two workshops. The first was conducted by Ecuador's Ministry of the Environment (MAE) to develop the *Action Plan for REDD+ and RIA*, which I attended along with representatives of COICA and other IPOs from Ecuador. The second was a communications and media workshop, which was held at the office in collaboration with the Ford Foundation a communications organisation named Burness. Both of these organisations came from the United States to facilitate training for members of COICA and other indigenous leaders in the build-up to COP21. These events have been *selected* because they broke away from the monotony of office life and provided moments where the observations I had begun to make were exemplified and amplified.

Liberty in the Andes

In a large conference room in the Universidad Andina Simón Bolívar, 30-something people in clothing typical of any other business meeting in Ecuador sat around a horseshoe of tables facing a large projector screen. The attendees were members of COICA, administrative assistants, and a few guests, including academic experts, a representative of the Ecuadorian Government, and me. At the opposite end of the room was an impressive painting that covered half of the wall – *Liberty in the Andes*, by Otaveleño artist Jorge Perugachi. The painting depicted Libertas, the Roman goddess, in the Andean mountains, with figures representing Indigenous Peoples' surrounding her and clinging to her waist,[12] but nobody seemed to notice it.

The first ever meeting of COICA had been held in Lima on March 14, 1984, and the meeting at the University was to mark the end of a five-day conference to celebrate its achievements and to plan for its future. After a morning of introductions and presentations about the history and ongoing work of the organisation, we chatted among ourselves or tapped away on phones while we waited for the food to be prepared and for the entertainment to begin. People took pictures of themselves and each other with the banner at the front of the room, *30 años de COICA* (30 years of COICA), and some uploaded them to Facebook, Instagram, and Twitter.

*

The first four days of the week had been held in the COICA office, and although most of the attendees were from IPOs, there were also visits from a number of non-indigenous actors, including partner organisations such as representatives of The Nature Conservancy, The Ford Foundation, Conservation International, and the EDF.[13] The president of the organisation, had opened the proceedings, outlining the objective of the week; the institutional strengthening of COICA:

> We are in the process of becoming stronger […] we now have greater visibility than ever, and an enhanced ability to work with international organisations.

He thanked everyone for attending, and then began the round of introductions. Each of the 30 people in the overcrowded room stated their name and where they were from. All nine Amazon countries were represented, and some people took the opportunity to comment on what they hoped to achieve during the week. The representative from Suriname, who was also the Coordinator for Women and Families,[14] addressed the room in English:

> We now have thirty years' experience, and we need to use this experience to move forward. We need to be more inclusive with more

> representation. It needs to represent *all* of the Amazon Basin. Let us ensure that the work we do here is not just talk, but it carries us forward. One of the problems that needs to be addressed is having multilingual people here in the offices of COICA.

Another person, a man from Guyana, also spoke in English, and after introducing himself requested that the issue of funding allocation be added to the agenda, which he said had been created before they arrived, and that member country groups had not been properly consulted. The President gave a brief response, stating it could be addressed as part of the existing agenda, and then the round of introductions continued, followed by a summary presentation of the organisation's future aspirations: *The Amazon Indigenous Agenda.*

Throughout the day and as the week moved forward, various aspects of COICA operations, its plans, and the challenges it faced were discussed, but the issue of representation, and in particular communication within the organisation, was a sticking point. I would like this to be replaced with the following text: During the discussion around health, for example, an issue was raised, once again by the representative from Suriname, who stated that the organisation acts as if it represents only the Spanish speaking countries. The question was deflected by the Chair, who said "we need to follow the agenda, we are speaking about health", but the President of the Confederation of Indigenous Nationalities of the Ecuadorian Amazon (CONFENIAE), sustained the point:

> If we cannot share our thoughts and feelings and culture, we cannot call ourselves democratic.

This was further supported by the representative from Guyana, who again spoke in English, translated into Spanish by the office receptionist:

> Communication is the key to any success. We are empowered within the organisation to get translation services, and we cannot wait for this until our Congress in 2017. If that is the case then we have no place here, we are just side-lined. I believe it is an administrative issue and that instead of saying we will discuss it later; we must deal with it now. With all the money we have, and all the projects that exist, without communication it won't work.

Once again, a brief response was given by the Chair, who said it is an important issue but urged the room to focus on the agenda.[15] Returning to the subject of health, a member of the Interethnic Association for the

Development of the Peruvian Forest (AIDESEP) raised the issue of race and discrimination in medical access, stating that:

> It is not a medical problem; it is a political problem. What many of the academics do not understand is that it is the result of bureaucratic racism.

This was supported by the president of CONFENIAE who added that:

> We need to work with the World Health Organisation and explain to them that we are afraid of vaccinations, and we need to revive our use of traditional medicines. When my son was sick with diarrhoea last week, I did not feel safe taking him to the hospital.

The representative from Guyana further added that:

> They [miners and loggers] bring diseases with them. Sharing education and communication is necessary because we need to learn from each other. The Government just wants to get rid of us because they want access to our land and to our resources. We can't wait for them; we have to make health our priority and we can have strength in numbers.

The number and diversity of issues on the agenda made for an exhausting and overwhelming experience. Each of the first three days lasted more than 10 hours, and it was difficult to keep track of all that was discussed, but some issues were more divisive than others. A particular moment that stood out was when the Coordinator of Women and Families, insisted that she be allowed to present the findings of a report that she had worked on over the past year. This was resisted at first, particularly by the President who said, "we have this problem every time, we need to stick to the programme", but eventually she was permitted to speak. As she presented, the noise in the room gradually increased from occasional whispers to a constant murmur, until it became difficult to hear what was being said. She was suggesting that the best examples of goods produced by Indigenous women could be displayed on the COICA website, and that these could then be brought to COP21 and used to promote the economic empowerment of Indigenous women. The President of COICA had left the room by now, apparently not interested in what she had to say, and once her presentation was over, she sat looking frustrated with what had happened. As the room quietened down, the man from Guyana stood and said:

> I personally find it offensive that the coordinator of COICA left the room to smoke a cigarette while the Women's coordinator was talking about the subject of women and families.

The President did not respond, but others in the room nodded in agreement.

The end of the third day, and much of the fourth, focussed on the lessons from COP20 and plans for COP21. The discussion centred around financing participation at the event, securing visas and accreditation for the conference, and the plans for the Indigenous Pavilion, which they hoped would become a permanent event at COP. A concern that was raised and agreed upon by most who had attended COP20 was that the Pavilion in 2014 had been so far from the main event site that it had been logistically very difficult to travel between them. The result was that attendance (though they estimated 20,000 people had visited) was low, and therefore, the visibility of COICA and Indigenous Peoples had been limited. It was agreed that addressing this was imperative. The second point that was agreed upon was that COICA and the Indigenous representatives who would attend COP21 would benefit from being trained on what to say, so that they could speak in a united voice. The representative from The Ford Foundation offered to help facilitate this, which eventually resulted in the workshop discussed later in this chapter.

*

For the final day, the conference was moved to the room in the University with the painting of "Liberty in the Andes", where the proceedings of the week would be summarised, and the Resolutions would be presented. The day began with the President of CONFENIAE introducing himself alongside the President of COICA. He did so in a language I did not understand, but many of the Other people in the room did. He, like everyone else, was wearing typical business attire, but he also wore a brightly coloured headband, as he had done all week. The language he used and the headband he wore distinguished him from the President who introduced himself in Spanish and wore no obvious markers of indigenous culture. The only thing remarkable about his style of dress was how similar it was to the other men in the room, and to the original founding members of the organisation.

Following the standard round of introductions, the people in the room were guided through a brief history of COICA, which began with a picture of the founders; nine young men, all with dark skin and dark hair, wearing shirts and jeans or chinos, and smart shoes. Though the current President and the other men in the room looked a little older, and although there were now women present (almost a third of the people in the room), the image was similarly business like, and did not look particularly *Indigenous*. The continuity of COICA's image over 30 years – professional, and progressive rather than exotic or traditional – was strengthened by the presence in the room of one of the men in the picture. He came up to stand in front of the projected image of himself. I was struck by how little he had changed since the photograph was taken. Though he had aged accordingly, his clothes were almost identical, his hair was the same, and one person shouted out, to everyone's amusement, that he had "only gotten shorter!".

The most notably indigenous objects in the room, besides the headbands, and the beaded medallions[16] worn by a few of the men in place of neckties (subtle, but significant markers of indigeneity), were the pieces of embroidered Andean fabric. They were framed and mounted on the walls, but they were not put there by COICA, and instead were part of a series of displays of preserved indigenous artefacts exhibited throughout the halls and rooms of the University.[17] It created a sense of dissonance. As Dean and Leibsohn (2003, pp. 14–15) note, this type of display both obscures "the radical transformations of the lives of indigenous peoples brought about as a result of colonization [... and betrays ...] desires to freeze indigenous people in the past, turning them (or aspects of their lives) into artefacts or relics of a bygone, romanticized era". And yet the room itself was full of Indigenous Peoples: "modern", "professional", and very much in the present.

Shortly before lunch, two women in more "traditional" Amazonian clothing (simple dresses – one blue, one red – colourful beaded necklaces, and brightly coloured feather earrings) also joined the room, but they sat separately at the back and watched the remainder of the presentations in silence. It had been announced that there would be a performance to end the event before lunch, and while waiting for it to begin, the men took turns photographing the *Indigenous* women, and having their picture taken with them. The women were contestants in the First Indigenous Beauty Pageant of Ecuador (*Belleza Indígena*). One of them went on to win the pageant, and though she was wearing clothing typically associated with the Amazon while at the COICA event, I was later told by one of the Indigenous men that she was, in fact, of Puruhá origin, a people associated with the Andean mountain province of Chimborazo.[18]

When the entertainment finally began, a group of eight young people, possibly in their late teens, walked into the room, slowly and ceremoniously (but smiling and holding back giggles) while a soundtrack of the forest played through a laptop. I wondered if it was the same audio that had been played at the Pavilion at COP20. They were barefoot and wearing artificial animal skins, including tiger print, and they looked like they were made of polyester. The men wore golden paper headdresses and carried wooden spears, and one of the women, the central figure of the dance, wore a small, beaded top and a short, leopard-print skirt.[19] They were all being *Indigenous* in the way that someone who had never met an

Indigenous person might imagine an Amazon tribe, and they danced and chanted while the attendees of the conference watched silently and took pictures and videos. The whole performance was quite strange to me. The young men were a little awkward and seemed poorly prepared, and the stylised portrayal of indigenous culture appeared exaggerated and almost comical in juxtaposition to the professionalised people in the room.

As the dance ended, the performers held a pose that reminded me of the bronze sculpture at the Indigenous Pavilion the year before,[20] but here, rather than the apex of the Geodesic Dome, the woman reached to the white tiles of the suspended ceiling in the University conference room. Most of the people in the room clapped quietly and some laughed a little, while the dancers posed for photos. What struck me most was that everyone else in the room seemed to enjoy the display, while I felt extremely uncomfortable.[21] The dancers left as they had arrived, and then we broke for lunch where I was handed a memory stick with the Resolutions from the conference and asked to translate them to English by the next day.

To see the indigenous Amazon world being depicted in such a way is not uncommon in Ecuador. On many occasions, I had been subjected to (and had paid for) overtly contrived and obviously "inauthentic" displays of indigenous culture. This included being taken up the river by a local guide to an indigenous "museum" – which contained grass skirts, coconut bras, and replicas of shrunken heads – near a small forest town for a "healing and cleansing ritual". I declined to take part but watched cynically as a Shaman chanted and blew smoke (from a Marlboro cigarette) into the hair and faces of my friends while his assistant hit them across their backs and arms with a bunch of sticks. What made the dance I observed at the conference unusual, though, was that the people watching were not tourists but were themselves Indigenous. Moreover, they were representatives from nine countries and so were undoubtedly familiar with the complex and varied cultures of Amazonian Indigenous Peoples. Indeed, part of the reason for the conference was to celebrate the contribution the organisation had made *to* indigenous culture. On the schedule of the conference, the day was titled *Interculturality, States, and Indigenous Peoples for the Construction of a Healthy World*, which suggested a sensitivity to the idea that indigenous cultures across the Amazon, though intertwined, are rich and diverse.

*

A presentation earlier that day, given by a technical consultant from AIDESEP, had focussed on the role of Indigenous Peoples and cultures in protecting the rainforest, and had begun with a slide that highlighted the need for a "cosmovisión integral" (a comprehensive worldview) for

COICA that advocated a "unity between culture, identity, forests, and territories". It reiterated a number of ideas that had been discussed throughout the week: there are no Indigenous Peoples without territories, there are no territories or forests without Indigenous Peoples, there are no Indigenous Peoples without self-determination, and living a "vida plena" (full life) depends on maintaining unity among Indigenous Peoples. He emphasised the complexity and wisdom of indigenous culture, and how this "holistic" approach makes Indigenous Peoples uniquely capable of managing the forests in which they live. His presentation went on to discuss some of the concerns relating to the programme for Reducing Emissions from Deforestation and Forest Degradation (plus) (REDD+), including "carbon piracy" and loss of access to resources, but also highlighted the potential opportunity that REDD+ presented, specifically with regard to securing territorial rights.

For this reason, the technical consultant argued that rather than accepting or rejecting REDD+, there is a third option, which is REDD+ Indígena Amazónico (RIA), a major component of the organisations Amazon Indigenous Agenda. He used maps like the ones hung near the desk in my office, that were based on various data, collected by Indigenous Peoples and from satellite imaging and other technical means, to remind the room that Indigenous Peoples are far from simple caricatures (like those portrayed in the dance). What became apparent to me here, was that the presentation he conducted was almost identical to the one I had seen at COP20 (and would later see at COP21), despite this time being presented to a largely indigenous audience. The argument being made for a version of REDD+ that incorporates indigenous knowledges is that the places where Indigenous Peoples live are not deforested as quickly, and the data is there to prove it. The claim is that this is because Indigenous Peoples know and understand the forest in ways that "we" (i.e., non-indigenous people) do not. And so, securing territorial rights for Indigenous Peoples is necessary in order to protect the forest and to mitigate climate change.

Cosmovisions, Consultation, and Consent

On the edge of *Gringolandia*,[22] the conference room of Hotel Sebastián overlooks the metropolitan centre of Quito. There is an invisible line dividing the room in two. It separates the thirteen "experts" from the seven Indigenous representatives (there had been 11 the day before, but some had not returned for the second day). Light skin is separate from dark; there are ties and gold earrings on one side of the room, bead medallions and wooden jewellery on the other. Though the people from COICA look the same as they did at the 30th Anniversary, being in direct contact with the non-indigenous people, who represented the institutions of the State, made their small markers of indigeneity more visible.

It was the second day of a workshop run by the MAE, the objective of which was to:

> Inform and receive feedback from the representatives of the indigenous peoples and nationalities of the Ecuadorian Amazon on the REDD+ Action Plan, in order to identify areas of potential opportunities and synergies with the RIA proposal.

The first day had consisted of various PowerPoint presentations. The "experts" outlined the general concepts of REDD+, its current position on the international agenda, and Ecuador's National Action Plan. The Indigenous representatives discussed their priorities and concerns in relation to REDD+ and outlined the concept of RIA. The main priority, from the perspective of the Indigenous people in the room, was to ensure that their rights and territories were protected, while their biggest concern was that REDD+ and market-based approaches could undermine this. As one member of COICA repeated a number of times throughout the workshop, territory and rights are, from the perspective of COICA, central to these discussions:

> Half of the carbon in the Amazon is in indigenous territories [...] as long as REDD+ continues to create reasons to discuss our land and our rights, it is an opportunity.

But as the first day of the workshop progressed, it became increasingly apparent to me that the question was limited to whether RIA, or Indigenous Peoples' views more generally, can be integrated or assimilated into a structure that already exists: the question that came up again and again was how, and indeed if, RIA can and should be included in the National REDD+ Action Plan. The discussion was never really about whether the Action Plan itself, or REDD+, were effective (or what this would even mean to various actors), or whether there might be better alternatives for Indigenous Peoples. It was never about integrating REDD+ into existing indigenous cosmovisions.

In the afternoon of the first day, the room had broken into groups for discussions and to plan presentations for the next morning. I had eaten lunch with an Indigenous man named Vicente, not out of choice, but because it was the only seat left. He was sat alone, looking out of the window over the city: *Gringolandia*. He was wearing a shirt and a pair of trousers that were far too big for him, and he had a deep scar that ran over his right eyebrow, across the bridge of his nose, and down onto his cheekbone. His silence and his appearance intimidated me.[23] But once I was seated, he began telling me about his children (he had ten) and his grandchildren. He was 57 years old, Kichwa, and part of the Federation of Indian Organisations of the Napo (FOIN).[24] Neither of us liked the food they had served and so we only ate the rice. He had said very little so far during the workshop, but when I asked what he thought of it, he told me:

> They don't understand how to involve us in this. They think that indigenous cosmovisions are *one* way of seeing the world, but they are many. The cities and development can be good for us, but our young people get lost in them, losing their culture, traditions, and families, and we never get what is promised: health, education, and development.

On the second morning, Vicente was sat next to me again, this time in the conference room. The presentations that had begun the day had largely consisted of the Indigenous people in the room reiterating what they had "learnt" the day before about the National Plan and how REDD+ works. The non-indigenous people spent no time demonstrating that *they* understood RIA or indigenous cosmovisions. As the conversations about the presentations began, Vicente stood and addressed the people on the other side of the invisible line. He appeared somewhat reluctant to start speaking, or unsure of what he was about to say, and he took an uncomfortably long moment to compose himself, looking down at a small piece of paper on the table in front of him. He then began. He spoke clearly, and slowly, but did not look up:

> The plans we are talking about don't include indigenous cosmovisions. They can't because when we are here even indigenous proposals have an *ojo occidentál* [Western perspective], because when we are here in *your* territory, in this hotel in the city, in this workshop, we are *in* your cosmovision.

Another Indigenous man responded immediately:

> *¡Sí, sí, sí!* Next time the workshop should be in the Amazon, in *our* territory, with *guayusa*, *yuca*, and *chicha!*

He was half joking, and everyone laughed, though the Indigenous people seemed to find it funnier than the non-indigenous people did. The "joke" and the preceding statement reasserted points that had been made the day before. Another Indigenous delegate from CONFENIAE had put forward a proposal for a regional-international REDD+ workshop with Indigenous Peoples and other actors, only this time to be conducted in indigenous Amazon territory in the manner of *their* cultural practices (though he had not suggested the *guayusa*, *yuca*, and *chicha*). The non-indigenous people in the room had listened and nodded along, but they had committed to nothing.

The implied ideological boundaries alluded to by Vicente had also been made more explicit in presentations during the first day. For example, when describing the "Normative, Political and Institutional Framework of the REDD+ Action Plan", a representative of the United Nations Development Programme (UNDP) stated that, "REDD+ isn't going anywhere. The decision has been made", and any conservation project that wishes to gain funding through it must be integrated into the National Programme. She pointed out that, although the vision on how to do so had changed over time, the aim is still to reduce greenhouse gas emissions, and so it is imperative that outputs can be measured in tCO_2e. She also noted that Article 74 of the Constitution of Ecuador states that:

> Persons, communities, peoples and nationalities will have the right to benefit from the environment and the natural riches that allow them to live well (*Buen Vivir*).

But also says that:

> Environmental services will not be susceptible to appropriation; its production, provision, exploitation and use will be regulated by the State.

The choice to cite this Article demonstrates the relative power of the non-indigenous actors in the room in relation to Indigenous Peoples: they define the terms of engagement. Quoting the constitution in this way, makes it very clear: these are the rules, and if you want to be part of these discussions, you have to follow them. It brought to mind a comment made by an Indigenous man from Panama I had met at the Indigenous Pavilion in Lima "When they want your resources, the first thing they do is change the law". It also highlights the asymmetry between the State and Indigenous Peoples: though the State grants Indigenous Peoples the "right" to benefit from the natural resources found in their territories, the resources and the "environmental services" are regulated and are *owned* by the State. There is no option other than to work within this legal framework as "good subjects".

Concerns raised regarding the underlying mechanisms of REDD+, particularly the role of the market, were brushed to one side. For example, when a member of COICA said that Indigenous Peoples need stable and ongoing funding, and that he was concerned about the potential risks of fluctuations in funding from market-based approaches to rainforest management, a representative of the MAE responded, stating that:

> REDD+ doesn't depend on markets because there are no carbon markets. It depends on donor funds. But donations also exist in a market so if you want to access funds you have to be competitive.

Though the encounter was intended to "Inform and receive feedback from the representatives of the indigenous peoples", it appeared that there was a definite limit to what could, and could not, be discussed. The non-indigenous people in the room expressed sympathy for the situation of Indigenous Peoples, and my impression was that there was a genuine motivation on a personal level to take their concerns into consideration. But it ended there: any concerns, critiques, or "alternatives" were relevant only if they could be understood within the epistemological boundaries, the "cosmovision", from which REDD+ emanates.

What seemed particularly difficult to get past was the contradiction that, although the UN recognises in principle that REDD+ programmes or projects in indigenous territories should have the free, prior, informed *consent* (FPIC) of Indigenous Peoples (Anderson, 2011, p. 1), the Cancún Safeguards, which were put in place in 2008 to minimise negative social and environmental effects REDD+, make no specific reference to FPIC. Nonetheless, the workshop was ostensibly an exercise aimed at informing and gaining consent from representatives of Indigenous Peoples, but those in the room had many concerns relating to the very concept of REDD+ and were offering an alternative. Yet, because that alternative did not fit neatly within the existing institutional framework of REDD+, which had already "been decided", it could only be included if it was twisted and distorted until it did. Both sides seemed to become frustrated with this impasse, and the conversations became cyclical.

I noticed that throughout the workshop, the terms "consultation" and "consent" were used interchangeably (FPIC and "consulta previa"), with no attention paid to the difference in meaning.[25] The conflation of the terms was a telling sign, and as the title of the workshop noted, the aim was to "inform", "receive feedback", and "identify synergies and opportunities": it was about consultation, not consent. This reminded me of something a member of COICA had said to me during our first interview some months before when I asked about why the safeguards of REDD+ were not enough; "It's like, if you have a Coca-Cola or a Coca-Cola Life.

You still have Coca-Cola". I found this comment revealing. Coca-Cola Life was introduced the year before and was sold in green cans, and its slightly reduced sugar content meant it could be marketed as a "healthier" alternative, and the marketing move was criticised as an attempt at "greenwashing" (Han, 2015). That this example was used in relation to REDD+ demonstrated to me that even those representatives of Indigenous Peoples who are working within its framework as apparently "good subjects" are under no illusion regarding what it really represents.

Teaching the Subaltern to Speak?

> Maintain good eye contact and an open posture. Be careful with defensive non-verbal signals. Keep them to a minimum. Don't cross your arms, and don't frown.

A white woman from the United States stood at the front of the room, reading from a booklet in Spanish. She was from a communications firm, Burness, which focusses on working with the non-profit sector.[26] Each person in the room had a copy of the booklet to take home with them. Most were Indigenous and worked for COICA or partner organisations, but there were also a few Others too: another member of Burness, a woman from The Ford Foundation, a man from the EDF, and me. Most of the people in attendance would also be attending COP21 and would be at the Indigenous Pavilion in Paris.

> Apologize when you have made a mistake, and say what is being done to correct it, and move on to the next topic. However, avoid apologizing too much.

I am bored. My mind wandered and I thought back to the sales training I had when I worked in a call-centre some years before. I found it tedious and patronising then, and I was struggling to take this "training" seriously. I was aware that I was projecting onto what I was observing my own experience of being told how I should speak and act, but it was hard not to: it was practically identical. Though the Indigenous people in the room were engaging with the tasks and listening to the presentations, I found it hard to imagine that they did not find it a little condescending too. Some looked, to me, as bored as I felt.

> Accept the frustration and anger of people, for example: "I can really understand how disgusted everyone is about this matter, we are doing everything possible to ..." Establish empathy ties.

Surely that has just been copied and pasted from a customer services manual? The woman is reading through the points quickly, summarising some of the ideas that came up during a role play that we had just done. First, we had sat in groups and listed our ideas of what makes a good interviewee. Then a camera had been set up at the front of the room to conduct mock interviews. The first man had sat in front of the camera. "Don't cross your arms!" someone called out, and everyone laughed. He hung his arms awkwardly at his sides, everyone laughed again, and then he shuffled in his seat, placed his hands on his knees, sat up right, and smiled. It did not look natural at all. The questions began.

> "Why are you here at COP21?" The interviewer asked.
>
> I am here to share the vision of my people for our future. We have a long history of living in harmony with the planet, but our lands are under threat from deforestation. Our territory is rich and beautiful, and protecting it is an important part of facing climate change. We have knowledge of the rainforest and our relationship with *Pacha Mama* can help to protect the planet from destruction. To defend our forests, our rights to our lives, and to our culture, our territory must be protected.

In the earlier part of the day, answering questions like these had been covered in the training:

> The message is the crux of the matter – the reason *why someone would be interested* [emphasis added] in your study, initiative, or announcement. The message provides the general image by providing context, a sense of urgency and/or possible steps to follow.

There had been a discussion that had focussed on what the message of Indigenous Peoples should be at COP21. It was during that session that the idea of "Paddle to Paris", the slogan used by numerous IPOs at COP21 was agreed upon by COICA. It cleverly merged the popular phrase "Road to Paris", which had become popular since COP20, with the plan that had been made by the people from Sarayaku (who were also represented at the workshop) to build and transport a Canoe from than Amazon to Paris. This and a new global carbon map of indigenous territories that, which would be unveiled at COP21, became central themes in the social media campaigns facilitated by Burness in the build up to and during the conference under the hashtags #PaddleToParis and #ifnotusthenwho. As reported on the Burness website:

> Divided by differences of language and culture, leaders representing millions of forest peoples on four continents had found common ground

> after several years of negotiations. Their efforts culminated in Paris with the joint release of an *evidence-based carbon map that strengthened forest peoples' claim to be an existing solution to climate change* [emphasis added].
>
> *(Sirica, 2015)*

The answer given in the mock interview had provided a good summary of the key points discussed: that Indigenous Peoples and their land was under threat, that their lives depended on their rights being upheld, and that their knowledge was uniquely "valuable" in combating climate change. And it had done so in a way that fulfilled the guidelines of the training, which explained that "bridging" should be used to return to "the message", regardless of the question. The example given by the trainer was from a 2007 United States Presidential candidate debate. The moderator, Tim Russert (following on from an obscure question directed at Dennis Kucinich regarding his claim that he had seen a UFO), addressed Barrack Obama:

> "I'm going to ask Senator Obama a question, in the same line. The three astronauts of Apollo 11 who went to the moon back in 1969, all said that they believe there is life beyond Earth. Do you agree?" Russert asked.
>
> You know, I don't know. And I don't presume to know. What I know is there is life here on Earth. And, and that we're not attending to life here on Earth. We're not taking care of kids who are alive and unfortunately are not getting healthcare. We're not taking care of senior citizens who are alive and are seeing their heating prices go up. So, as President, those are the people I will be attending to first.

Obama's response demonstrated mastery of the art of "bridging". It is an effective tool, and it is characteristic of Western political discourse: people do not remember the question, only the answer. It occurred to me that what the people here were learning to do was something similar to what I had observed being done to *them* at the meeting with MAE earlier in the month. But the same man who was seated in front of the camera – smiling, his hands placed politely on his knees – gave me a different message when we spoke at lunchtime. I asked what he would really like to say at COP21, and he leaned in, with a smile on his face, and said:

> If I could, I would take the shrunken heads of our past enemies to the UN, hold them up and tell them, if you come to try to take our land, then this is what we will do to you.

He laughed and nudged me with his elbow, and we continued eating. I knew he was joking, but there was an element of truth in what he was saying.

It echoed things I had been told by Indigenous people on various occasions, and declarations made by IPOs that took a stronger tone. For example, at the Cumbre de Los Pueblos, an event that happened alongside COP20 in Lima, Manari had said:

> They think they are the gods, but they only have the power that we give them. Our position is firm, and we have no fear, and with that strength we will continue. Climate change and conservation are a new form of colonisation of our territory. The forests belong to us, and our elders have taught us to govern ourselves. We do not have to ask permission for that. We are not going to give up being the owners of our land, as we have always been.
>
> *(Manari, 2014)*[27]

At the end of the training while the facilitators were packing up, I asked the woman from the Ford Foundation how she thought the training had gone. She said it had been "good fun", and when I asked her opinion on Indigenous Peoples using the language of "sustainable development" and "valuing rainforests" in a financial sense, and particularly the choice of the term REDD+ in the name of their alternative proposal, RIA, she said "I don't like it at all, it's a terrible idea, so confusing". I agreed. Despite the time spent in the office and discussing it at length with COICA, I had found it difficult to understand how RIA was unique and different from REDD+. But I also saw the utility of the name, and so I asked her "but would you even be here if they weren't aligning themselves with these kinds of ideas". She sighed, and said "No, you're right, probably not".

Conclusion

Indigenous people using concepts that originate in Western discourse, including notions of indigeneity, and working with and within development institutions (such as the UN and "Big Green" NGOs) in an attempt to confront problems that are largely the result of Western domination is far from a new phenomenon. Robins (2008), for example, observed the ways in which entire communities have selectively reconstituted themselves as both "modern" and "traditional" in order to gain funding from NGOs. This has allowed them to benefit from development projects, and it has also contributed to the emergence of indigenous modernities: "by participating in NGO and donor driven projects, indigenous groups such as the Kalahari San are drawing on the modern institutions and resources of a global civil society to reconstitute themselves as a traditional community" (Robins, 2008, pp. 63–64). A less than generous reading of this process could lead to cynical and essentialist critiques relating to "authenticity" and legitimacy, and COICA has been subject to both. This is precisely the "double-bind" described by Baker et al. (2016, p. 2), which puts Indigenous people who seek to interact with and benefit from "sustainable development" in an almost impossible position. Mookherjee (2010) and Dion-Buffalo and Mohawk (1992) also described a version of this problem: either work within the system and be "good subjects" making compromises in order to benefit from development interventions, or do not compromise, be "bad subjects", and and risk being excluded from them.

COICA takes the first of these options: to gain a foothold within the development institutions of the Global North, its compromises have been numerous, going far beyond reconstituting communities in terms of indigenous modernities. Instead, they have had to leave their communities, in both the physical and cultural sense, existing instead on the very edge of Western notions of indigeneity. The (re)presentation of indigeneity among members of COICA, such as the dance performance and the presence of the Indigenous beauty pageant contestants at the 30th Anniversary appeared to me to be an active process through which the notion of being *Indigenous*, not to anywhere in particular, but in contrast to being non-indigenous, was being "made up" and brought into existence in similar ways to those I later observed at COP21. But it was different because it was happening *among* a group of Indigenous people, rather than for the benefit of Others.

Recreating and maintaining a coherent and shared identity is a crucial part of COICA's ability to perform their role as representatives and to act legitimately on behalf of *all* Indigenous Peoples from the Amazon. But doing so has also required the organisation and the people who work within it to take on specific elements of an imagined indigeneity – ecological nobility, the stewardship role, and the idea that there is "value" in indigenous knowledge – while also demonstrating a willingness and ability to discuss *their* territories, and to negotiate *their* "rights" within the Western epistemological framework: i.e., in terms of evidence and providing a reason *why* someone would be interested in what they have to say. This means measuring and mapping forests in terms of tCO_2e, discussing and measuring knowledge in terms of "contributions" to climate change, and adopting modes of communication that make this intelligible and "valuable" to non-indigenous people. They have become very "good subjects" indeed. And yet, a friction remains. This was particularly visible in the workshop with the Government of Ecuador and the UN, where it became clear that, despite supposedly attempting to incorporate Indigenous Peoples' concerns and approaches into the national REDD+ readiness programme, consultation did not necessarily lead to meaningful inclusion or consent. The discussion was strictly limited to how and why Indigenous Peoples' knowledge can and should be incorporated into the national approach. The conversation never considered if the national approach could be, or should be, incorporated into Indigenous Peoples' existing ways of being, even if the "evidence" suggests they had "worked" so far.

Working with organisations like Burness to translate and disseminate their "message" is effective at getting these points over in forums like COP, and it also contributes to building relationships with major funders, particularly those who are described by Klein as "Big Green" and "pro-trade" environmental NGOs (Klein, 2014, p. 84), such as the Nature Conservancy and EDF. Though I do not know how the partnership between COICA and Burness came about, there are two possibilities: either Burness chose COICA as an organisation that represents the type of "extraordinary people" whose stories they are committed to disseminating, or COICA, with its support from the "Big Green" NGOs, chose to hire Burness to help raise their profile. Both possibilities demonstrate the process of selection whereby those Indigenous people who work most effectively with and within the discourses and institutional logics of non-indigenous organisations become the most visible and legitimised representation *of* indigeneity in discussions around climate change mitigation. And this notion of indigeneity becomes what it is to be *Indigenous*. The presence and prominence of COICA at COP is dependent on this, and in that sense, it is a successful strategy, but it is also functional to certain non-indigenous interests. What is important to note here is that while the process of selection is happening in both directions, it is not an equitable interaction. Governments, NGOs, and other major funders, like the UN and the World Bank (all of whom COICA have worked with over the years), tend to speak in terms of "climate change mitigation" and "sustainable development", but for the people COICA represents it is the territory, resources, cultural survival, and the lives of Indigenous Peoples that are being negotiated, as opposed abstracted tonnes of carbon dioxide.

The relational encounters between epistemological communities in these spaces shape and are shaped by such systemic asymmetries of power. The limits of what can and cannot be discussed that are agreed upon either implicitly or explicitly (as demonstrated in the workshops with the Government of Ecuador and UN representatives, as well as the concerted effort to develop a media strategy for COP), require COICA to "translate" its message into languages and ideas that are intelligible and permissible within Western discourses around climate change and indigeneity. The same is not true in the opposite direction. Throughout my time at COICA and at COP, I saw no significant effort on the part of non-indigenous actors to engage with Indigenous Peoples *on their terms, using their languages or epistemologies, in their spaces*. The requirement for Indigenous Peoples to translate their knowledge and to incorporate Western discourses into their plans for their future in order to be afforded the "right" to self-determination represents a form of epistemological subordination of indigenous cosmovisions.

Thus, in pursuing and achieving representation within the remnant structures of a colonial past (such as the UN), COICA goes some way towards securing a degree of agency with regard to the rights and cultural survival of Indigenous Peoples. But this also fulfils the desire of the colonial project(s) "to be embraced, to be loved against all of the accumulated evidence pointing in the opposite direction" (Depelchin, 2005, p. 80), which in turn legitimises, reinforces, and recreates the imaginary Indians of green development fantasies (Tsing, 2007). "They" are still "savage" enough to the occidental eye to act as a sufficient representation of the Other but are not so antagonistic that they represent a challenge to existing power structures or epistemological hierarchies.

As will be made clear in the chapter that follows, my presence at COICA was itself the result of this process of selection. The organisation and the people did not represent the type of indigeneity I had imagined I would be writing about, or the type of people with whom I had intended to spend time. But their openness and willingness to spend time with *me*, their professionalism, and their geographical and cultural location meant that doing research about them, and therefore writing about them, was more pragmatic than working with other Others. COICA offers neither a simple reflection of a Western fantasy of an ideal Other, nor does it entirely reject or subvert such imaginaries and become a bad subject. In this sense, the organisation and its members occupy a precarious position, vulnerable to critiques of its legitimacy and to accusations of inauthenticity, as the Optimal Other. This ability to bridge worlds, to separate itself from certain elements of indigeneity and to move in spaces that are culturally and geographically distant from most Indigenous Peoples could be seen as the organisation's greatest strength and also its greatest weakness.

Notes

1 The very name of the Amazon region originates in Orellana's "discovery". After seeing the native women fight alongside the men, Orellana likened them to the Amazons of ancient times whose warrior women nourished "savage monsters" (minotaurs and centaurs) at their breast, and then fought alongside them in battle. He named the region and the people he "discovered" with this in mind, making permanent the connection between the place and the people and notions of savagery (McClure, 2017, p. 56).

2 Middleton (2009, p. 200) notes that, as with other colonial capitals, Quito's beauty, particularly in the Historic Centre, is a physical manifestation of the

historic concentration of power and wealth in Ecuadorian society, as well as being emblematic of half a century of exploitation of the indigenous population.

3 These violent clashes are at times viewed as part of ongoing State repression, intimidation, and racism against indigenous people (see e.g., Lu, Silva, & Valdivia, 2016; Martínez Novo, 2018), and though the country has what is deemed to be a progressive Constitution (Becker, 2011, p. 47), the *Freedom in the World* country report stated that there "remained a hostile environment for freedom of expression in 2016" and that "the government has a poor record regarding respect for civil liberties, particularly freedom of expression" (Freedom House, 2017).

4 CONAIE of Ecuador, and particularly *¡Pachakutik!* (often referred to as the "political arm" of CONAIE), for example, has historically been and remains (despite recent instability) the most visible and active indigenous political entity in the country (see e.g., Beck & Mijeski, 2011).

5 See e.g., *Rethinking indigenous politics in the era of the "indio permitido"* (Hale, 2004) and *Development or Decolonization in the Andes* (Apffel-Marglin, 1995).

6 As noted in the previous chapter a parallel can be drawn here to some understandings of shamanism, which allows, and sometimes *requires*, a person to become separate to and different from their community.

7 It is important to note that I do not intend to make a value judgment regarding the changes that occur during this process as a "loss" in terms of identity, or as a "gain" in terms of "modernization", or even to suggest that it is a neutral form of "hybridization". I only seek to illustrate how *they* appeared to *me* in this context. In the past, as Warren noted, anthropological discussions regarding how "indigenous culture" changes have almost invariably been framed in terms of impoverishment of culture or as loss. The challenge is to *both* acknowledge the colonial domination of indigenous people and to recognise the role of indigenous people in transforming their own culture and identity (Warren, 1978, p. 6).

8 As MacClancy (2013, p. 169) notes, the experience of feeling "bored" is not unique to an anthropologist confined to an office of this kind in a metropolitan city. Fieldwork anywhere can be "emotionally intense, psychologically enriching, potentially traumatic and, possibly dangerous [...but...] where extremes are exposed and may even collide, the worst sin of an anthropologist is to find all this less than engrossing. The fieldworker's heresy? To think the locals boring". MacClancy cites Rush who noted that: "All over the world in the privacy of their huts anthropologists are turning up their hands and saying: This is Boring. Life should not be boring" (Rush, 1991, p. 143, cited in MacClancy, 2013, p. 169).

9 The RAISG is a network that includes the Instituto del Bien Común, Instituto Socioambiental, Imazon, Gaia Amazonas, Provita, Eco Ciencia, Fundación de la Naturaleza, and is supported by NORAD (The Norwegian Agency for Development Cooperation), the Gordon and Betty Moor Foundation, and Regnskogfondet (Rainforest Foundation of Norway) (RAISG, 2012).

10 At the time I did not know what it meant, but *Pachakuti* has two meanings. The first, in Ecuador specifically, is that it is related to the name of the "political arm" of CONAIE (the national indigenous confederation) Pachakutik, which emerged as a "third option" for indigenous political activity that brought together indigenous and other popular movements to work together as equals on common goals, including "democratisation of the market" and creating equal opportunities for the most vulnerable groups in society as a political *movement*

not a political *party* (Becker, 2011, p. 44). Such popular movements were powerful enough to overthrow three governments between 1997 and 2005, and Pachakutik would be instrumental in bringing President Rafael Correa to office, and in bringing *Sumak Kawsay* (which became *Buen Vivir*) into mainstream political discourse in the country (Becker, 2010, p. xi). But as Cusicanqui notes, the word *Pachakuti* carries complex meaning that is often overlooked "The contemporary experience commits us to the present – *aka pacha* – which in turn contains within it the seeds of the future that emerge from the depths of the past [...] The present is the setting for simultaneously modernizing and archaic impulses, of strategies to preserve the status quo and of others that signify renewal of the world: *Pachakuti*" (Cusicanqui, 2012, p. 96).

11 In 2014, the relationship between the Government of Ecuador and some indigenous organizations had deteriorated, eventually resulting in the Correa administration's attempt to shut down CONAIE's headquarters in Quito (the building had been provided by the state in 1991), allegedly because the organisation had been "doing too much politics" (Valdivia, 2015, p. 603). On December 19, 2014, through the United Nations Permanent Forum for Indigenous Peoples, CONAIE declared a "State of Emergency", claiming that the move represented "an act of political persecution against the Indigenous movement" (UNPFIP, 2014). In 2013, the Government of Ecuador had dissolved and shut down the offices of Funadación Pachamama (a long established environmental NGO) which the organisation described as "an arbitrary act that seeks to suppress our legitimate right to dissent from the decision of the National Government to concede areas of Amazonian indigenous nations to oil companies, without respecting their constitutional rights to free, prior and informed consultation" (Pachamama, 2013). This pattern continued with an unsuccessful attempt to shutdown Acción Ecológica, one of Ecuador's most prominent Environmental NGOs, in 2016, which representatives of the organization described as the latest in a long chain of "different types of aggression" stating that "This isn't just about defending nature or land, but the right to participate, the right to work together, the right to protest, the right to speak" (Hill, 2017).

12 The use of a European goddess as a symbol of liberation for indigenous people stood out to me in the context of the event and the history of Ecuador. It brought to mind Sawyer's analysis of the Guayasimin's mosaic in the Presidential Palace in Quito (2004, pp. 32–33) that depicts the journey of Orellana into the Amazon: a masterpiece filled with contradiction, celebrating both the courage of as well as the victory over the savage Indians of the Amazon, and simultaneously remembering "the sacrifice of three thousand aboriginals" which "glorifies the presence of Ecuador in the Amazon River. The route is marked by their blood in our spirit". Peruganchi's piece is equally contradictory and to me was illustrative of a broader narrative. The recognition of the oppression of indigenous people coupled with the Western symbolic representation of liberty was a fitting analogue for the notion that Indigenous Peoples can be liberated through Western led development institutions that, despite a shift toward indigenous participation and "representation" (see e.g., Muehlebach, 2001) and the emergence of "holistic" and "integral" language, remain bound to neoliberal principles and the continuance of Western modern/colonial imposition: "the very idea of development itself is a concept and word that does not exist in the cosmovisions, conceptual categories, and languages of indigenous communities" (Walsh, 2007, p. 17).

13 It is worth noting that these partner organizations are not simply benign, silent donors. They have significant business interests and have all been criticized for

various activities conducted in the name of "environmentalism". Klein (2018) cites numerous examples of the "compromises" that have been made by these "Big Green" NGOs including: the EDF has actively supported both fracking and carbon trading and received $64 million in funding from Walmart between 2009 and 2013 (pp. 209, 215–217 and 226–229), The Nature Conservancy has itself extracted oil on land that was donated by Mobil as part of the "Texas City Prairie Reserve" and has administered a carbon offset project that allegedly stopped indigenous Guarani in Brazil from foraging for wood or hunting in their territory (pp. 192 and 221), Conservation International has partnerships with Walmart, Monsanto, Shell, Chevron ExxonMobil, Toyota, MacDonald's and BP (p. 196), and The Ford Foundation holds $14 million in Shell and BP stocks alone (p. 198).

14 As Mila explained to me, the name of her role as the Coordinator for "Women and Families" was decided upon in favour of the term "Gender" because "that doesn't make sense to us, but we still have to use it sometimes" because it is what the NGOs and donors use. As Smith (1999, p. 48) noted, gender, along with its distinctions and hierarchies are: "deeply encoded in Western languages. It is impossible to speak without using this language, and, more significantly for Indigenous Peoples, it is impossible to translate or interpret our societies into [...them...] without making gendered distinctions". Lugones (2008, p. 31) adds to this that the coloniality of gender has historically been entangled with systems of control and power: "This gender system congeals as Europe advances the colonial project(s). It begins to take shape during the Spanish and Portuguese colonial adventures and becomes full blown in late modernity. The gender system has a "light" and a "dark" side.

15 The adherence to the agenda and rigidity of the structure of time stood out to me for two reasons. First, it was the opposite of my experiences when attempting to conduct workshops in an indigenous community (see following chapter), and second, it appeared to be being used in order to prioritise some voices and concerns over others.

16 O'Driscoll observed that these "medallion-style" necklaces, along with physical characteristics including "Chinese eyes", dark skin, short stature, and stocky build, are considered among the Shuar people to be markers of their identity, noting that the medallions are "particularly popular with political figures" (O'Driscoll, 2015, p. 185).

17 In both the Global North and the Global South, according to Andreotti, Bruce, Stein and Suša (2016, p. 14), universities are bound up with global flows of capital through industry, research partnerships, grant funding, student loan debt, and other avenues and as such are "firmly rooted in both the content and the frame of the dominant, modern/colonial global imaginary, including its commitments to Western-defined notions of progress, social mobility, and success, though they have somewhat different emphases". On the subject of these "different emphases", Cusicanqui (2012, p. 98) criticises North America approaches to subaltern studies in general, and specifically Walter Mignolo (who was named as a "Permanent Researcher at Large" at the Universidad Andina Simón Bolívar), for "creating a jargon, a conceptual apparatus, and forms of reference and counterreference that have isolated academic treatises from any obligation to or dialogue with insurgent social forces. Walter Mignolo and company have built a small empire within an empire, strategically appropriating the contributions of the subaltern studies school of India and the various Latin American variants of critical reflection on colonization and decolonization".

18 With the proliferation of anthropologists and other researchers (from both Ecuador and elsewhere) who focus on indigeneity in Ecuador, it is somewhat surprising that little research has been conducted into the role of these beauty pageants in the construction of indigenous identity. However, Rogers analysed the role of municipal indigenous and non-indigenous pageants in Ecuador and the way that indigeneity is integrated into them: "During an early phase of the [non-indigenous] competition, [white] participants dance in some approximation of traditional indigenous clothing, to traditional musical themes, using traditional dance steps. Yet this apparent assumption of indigenous identity is immediately shed, never to be revisited over the course of the pageant. Thus, while actual indigenous people are not present at the performance, indigenousness is acknowledged as playing some role in the constitution of a generalizable municipal identity. But this identity is discarded in favour of a transcendent whiteness that incorporates indigenousness as a constitutive but not characterizing element. The indigenous pageant, on the other hand, is executed as part of a larger effort on the part of indigenous political organizations to 'seize control of and separate out of the larger society' cultural elements that can be used to create an ideology of 'ethnic worth'" (Rogers, 1998, pp. 56–57).

19 That neither tigers nor leopards (particularly ones made of polyester) exist in the Amazon was apparently not an important detail.

20 As noted in the previous chapter, my disposition as a researcher became more and more awkward throughout my fieldwork, which resulted in seeing myself as both more "inside" and more "outside" of my research. On the one hand, I became increasingly aware of the role of anthropology and anthropologists (and research and academia in general) in continuing to uphold extant structural inequalities and the Othering of research "subjects". But conversely, my own experiences *of* my research (see subsequent chapters) affected my personal connection *to* my research, which resulted in me feeling very much "outside" of the experience: always observing, but never participating.

21 I had seen a number of performances like these at events organized by indigenous people, but the context in the University as well as the contrast between the dancers and the delegates made this one seem particularly "inauthentic", perhaps because universities tend to be more preoccupied with the preservation of "real" culture.

22 *Gringolandia* is the local term used for La Mariscal, a part of the city that has many shops, hotels, hostels, bars, and restaurants, as well as Quito's main artisan market. It is popular with tourists, immigrants (of the "expat" variety), and young metropolitan Ecuadorians, and is known for being very Westernised.

23 I was disappointed in myself for how readily I had internalised the idea that facial scars would tell me anything about a person, but it is a ubiquitous artistic trope, particularly in cinematic depictions of "villains" (see e.g., Croley, Reese and Wagner's [2017] analysis of the "hero-villain skin dichotomy"). But it was also a useful reminder of how my position is not impartial or neutral, and how much work is to be done in order to overcome even the most obvious stereotypes of "savages".

24 FOIN is part of CONFENIAE, which is a member of COICA.

25 The same inconsistent use of these two terms happened during the COICA 30th Anniversary Events. I was asked to provide the organisation with a summary of my observations, in which I raised the point. I was told that it was an ongoing issue. While FPIC is an important part of the UNDRIP (see Art. 10 UNDRIP, 2008, p. 6), the Constitution of Ecuador uses the term "consulta previa", prior consultation (GOE, 2008, pp. 41 and 179), which may be the source of the inconsistency in the use of the terms.

26 The Burness website states that the organisations "unwavering mission is to help extraordinary people tell their stories for the good of the world" and provides services including message development, strategic communications, and training workshops that focus on themes such as being memorable, controlling an interview, and messaging, to many non-profits (Burness, n.d.).
27 This particular quote is paraphrased from notes that were taken during the speech, but to the best of my knowledge are an accurate representation of what was said.

References

Anderson, P. (2011). *Forests and climate change after Cancun: An Asia-Pacific perspective*. Retrieved from https://redd.unfccc.int/uploads/63_7_redd_20110503_recoftc_after-cancun.pdf

Andreotti, V., Bruce, J., Stein, S., & Suša, R. (2016). Towards different conversations about the internationalization of higher education. *Comparative and International Education/Éducation Comparée et Internationale*, *45*(1), 2.

Apffel-Marglin, F. (1995). Development or decolonization in the Andes? *Futures*, 27(8), 869–882.

Baker, L., Brock, S., Dove, M. R., Hebdon, C., & Ludlow, F. (2016). The double binds of indigeneity and indigenous resistance. *Humanities*, *5*(3), 53.

Beck, S. H., & Mijeski, K. J. (2011). *Pachakutik and the rise and decline of the Ecuadorian indigenous movement*. Athens: Ohio University Press.

Becker, M. (2010). *Pachakutik: Indigenous movements and electoral politics in Ecuador*. Plymouth: Rowman & Littlefield.

Becker, M. (2011). Correa, indigenous movements, and the writing of a new constitution in Ecuador. *Latin American Perspectives*, *38*(1), 47–62.

Burness. (n.d.). *About*. Retrieved from https://www.burness.com/

Croley, J. A., Reese, V., & Wagner, R. F. (2017). Dermatologic features of classic movie villains: The face of evil. *JAMA Dermatology*, *153*(6), 559–564.

Cusicanqui, S. R. (2012). Ch'ixinakax utxiwa: A reflection on the practices and discourses of decolonization. *South Atlantic Quarterly*, *111*(1), 95–109.

Dean, C., & Leibsohn, D. (2003). Hybridity and its discontents: Considering visual culture in colonial Spanish America. *Colonial Latin American Review*, 12(1), 5–35.

Depelchin, J. (2005). *Silences in African history: Between the syndromes of discovery and abolition*. Dar es Salaam: Mkuki na Nyota Publishers.

Dion-Buffalo, Y., & Mohawk, J. C. (1992). Thoughts from an autochthonous center: Postmodernism and cultural studies. *Akweikon Journal*, 9(4), 16–21.

Freedom House. (2017). *Freedom in the world. Country report: Ecuador*. Retrieved from https://freedomhouse.org/country/ecuador/freedom-world/2017

GOE. (2008). *Constitución Del Ecuador*. Retrieved from http://www.wipo.int/edocs/lexdocs/laws/es/ec/ec030es.pdf

Hale, C. (2004). Rethinking indigenous politics in the era of the "Indio Permitido". *NACLA Report on the Americas*, *38*(2), 16–21.

Han, E. (2015, March 30). Coca-Cola accused of 'greenwashing' with launch of sugar-reduced Coke Life. *Sydney Morning Herald*. Retrieved from https://www.smh.com.au/healthcare/cocacola-accused-of-greenwashing-with-launch-of-sugarreduced-coke-life-20150330-1mas6u.html

Hill, D. (2017, January 21). Ecuador's leading environmental group fights to stop forced closure. *The Guardian*. Retrieved from https://www.theguardian.com/environment/andes-to-the-amazon/2017/jan/07/ecuadors-leading-environmental-group-fights-forced-closure

Klein, M. A. (2018). A critique of colonial rule: A response to Bruce Gilley. *Australasian Review of African Studies*, *39*(1), 39–52.

Klein, N. (2014). *This changes everything*. London: Random House.

Lu, F., Silva, N. L., & Valdivia, G. (2016). *Oil, revolution, and indigenous citizenship in Ecuadorian Amazonia*. London: Palgrave MacMillan.

Lugones, M. (2008). The coloniality of gender. In W. Harcourt (Ed.) (2016). *The Palgrave handbook of gender and development* (pp. 13–33). London: Palgrave Macmillan.

MacClancy, J. (2013). *Anthropology in the public arena: Historical and contemporary contexts*. Oxford: Wiley Blackwell.

Martínez Novo, C. (2018). Ventriloquism, racism and the politics of decoloniality in Ecuador. *Cultural Studies*, *32*(3), 389–413.

McClure, J. (2017). *The Franciscan invention of the new world*. London: Palgrave Macmillan.

Middleton, A. (2009). Trivialising culture, social conflict and heritage tourism in Quito. In A. Ypeij & M. Baud (Eds.), *Cultural tourism in Latin America* (pp. 199–216). Leiden: Brill.

Mookherjee, M. (2010). Postcolonial multiculturalism. In D. Ivison (Ed.), *The Ashgate research companion to multiculturalism* (pp. 179–198). London: Routledge.

Muehlebach, A. (2001). "Making place" at the United Nations: Indigenous cultural politics at the UN working group on indigenous populations. *Cultural Anthropology*, *16*(3), 415–448.

O'Driscoll, E. (2015). *Contested identities: Urbanisation and indigenous identity in the Ecuadorian Amazon* (Unpublished doctoral thesis, Department of Social Anthropology, University of Kent, Canterbury). Retrieved from https://kar.kent.ac.uk/47959/1/73PhD%20Thesis.pdf

Pachamama (2013). Fundación pachamama statement on shutdown of their offices. Retrieved from https://news.pachamama.org/news/fundacion-pachamama-on-international-and-public-opinion.

RAISG. (2012). *Amazonía 201: áreas naturales protegidas y los territorios indígenas*. Retrieved from https://www.amazoniasocioambiental.org/es/mapas/#!/descargas

Robins, S. L. (2008). *From revolution to rights in South Africa: Social movements, NGOs & popular politics after apartheid*. Rochester, NY: Boydell & Brewer.

Rogers, M. (1998). Spectacular bodies: Folklorization and the politics of identity in Ecuadorian beauty pageants. *The Journal of Latin American and Caribbean Anthropology*, *3*(2), 54–85.

Sawyer, S. (2004). *Crude chronicles: Indigenous politics, multinational oil, and neoliberalism in Ecuador*. London & Durham: Duke University Press.

Sirica, C. (2015). *Indigenous peoples 'Paddle to Paris' for COP21*. Retrieved from https://www.burness.com/indigenous-peoples-paddle-to-paris-for-cop21/

Smith, L. T. (1999). *Decolonizing methodologies: Research and indigenous peoples*. London: Zed Books.

Tsing, A. L. (2007). Becoming a tribal elder, and other fantasies of green development (reprint). In M. Dove & C. Carpenter (Eds.), *Environmental anthropology: A historical reader*. Malden, MA: Blackwell.
UNDRIP. (2008). *United Nations declaration on the rights of indigenous peoples*. Retrieved from https://www.un.org/esa/socdev/unpfii/documents/DRIPS_en.pdf
UNPFIP. (2014). *CONAIE: Declaration of emergency*. Retrieved from http://unpfip.blogspot.fr/2014/12/conaie-declaration-of-emergency_19.html
Valdivia, G. (2015, September 7). Review of the book *Governing indigenous territories: Enacting sovereignty in the Ecuadorian Amazon* by J. S. Erazo. *American Anthropologist*, *117*(3), 602–603.
Walsh, C. (2007). Shifting the geopolitics of critical knowledge: Decolonial thought and cultural studies 'others' in the Andes. *Cultural Studies*, *21*(2–3), 224–239.
Warren, K. B. (1978). *The symbolism of subordination: Indian identity in a Guatemalan town*. Austin, TX: University of Texas Press.

PART III

Fieldwork that Hurts

5

FAILURES AND FRUSTRATION DURING RESEARCH ENCOUNTERS

Introduction

My own journey into the Amazon began at 6.30 am on June 6, 2014. I walked down the hill, along the Camino de Orellana, and east toward the rainforest. I bought a cigarette and a Coca-Cola from the tienda opposite the Catholic Church, and sat in front of the statue of Francisco de Orellana, "Discoverer of the Amazon". I read the plaque again, but this time noticed that he had been about the same age as me when he began his Canoe Trip.[1] I did not feel as though we had much else in common, though. To me he was a symbol of all that was wrong with the history of Latin America. He represented colonisation, violence, and subjugation. On my adventure, by bus and by plane, rather than on foot and by river, I was going to the rainforest for a different reason. I was there to *help*.

Early on in my fieldwork in Ecuador, I was contacted by a friend who asked if I would be interested in working with a Peruvian NGO that was investigating how men and women in a small indigenous community were experiencing climate change, and what their perspectives were on the emerging REDD+ programme. It was part of an international study, working with Indigenous Peoples around the world, coordinated by an organisation called The Indigenous Peoples Biocultural Climate Change Assessment (IPCCA). Like the Indigenous Pavilion (which is abbreviated to the IIPFCC), its acronym mirrored that of the IPCC, immediately lending it an apparent legitimacy. My friend had been excited to tell me about the opportunity because, in her words, this community was "as real as it gets in Ecuador".

DOI: 10.4324/9781003341864-9

Based on extensive conversations we had had regarding the Indigenous Peoples in the country, and her work relating to carbon trading and colonialism, I took this to mean that the Sápara people were continuing to struggle, against the odds, to defend their culture and territory from extractivist, neo-colonialist policies and the risks associated with carbon markets. We had lamented the history of oil contamination, the destruction of indigenous territories, and the politicisation of the indigenous movement in Ecuador. And we had discussed our frustrations with the role environmentalism now appeared to be playing in continuing this trend. A few years before, I had spent time researching a lawsuit against Chevron for environmental contamination, and though I had visited the edge of the rainforest, the small number of Indigenous people I had met were accustomed to life in towns and cities, and many had travelled to Europe and the United States to speak on behalf of their communities. Gaining access to those Indigenous people who have a more antagonistic relationship with outside actors, those who choose *not* to assimilate, and *not* to negotiate, had so far proven to be challenging, and so I was excited to have the opportunity to at least initiate a relationship with such "real" Indigenous people.

This chapter reflects on my brief experience of working directly with the Sápara people, and what I learned in the build-up to doing so as I tried to find out as much as I could about their history and culture. My main finding was that few people, apart from the Sápara themselves, know very much about their lives. They live in one of the most bio-diverse regions of the world in a territory that straddles the border between Ecuador and Peru. Thought to have once numbered over 100,000 people, a history of conquest, slavery, disease epidemics, forced conversions to Christianity, war, and environmental damage (UNESCO, 2008) led anthropologists in the 1980s to believe that the Sápara people had been wiped out and their culture and language entirely lost. However, deep in the rainforest a small group survived and some still spoke the distinct Sápara language[2] (Curwood, 2004). In 2001 it was believed that the population had dwindled to only 176 people (Pallares, 2001, p. 12), and they are still among the smallest indigenous groups in the region. Yet their culture, rich in knowledge of the forest, based on an oral tradition expressed through myths, rituals, and art, and their language, with an extensive vocabulary for flora and fauna, has just about survived. UNESCO has recognised that this discrete knowledge and language "constitutes the memory of the entire region", declaring it a Masterpiece of the Intangible and Oral Heritage of Humanity (UNESCO, 2008).

Today, the number of Sápara people living in Ecuador is unclear. Estimates vary greatly ranging from 100 to 600, but UNESCO in 2014 estimated that the population had dwindled to no more than 300 in total: 100 in Peru and 200 in Ecuador (UNESCO, 2014). Part of the reason for the discrepancy is that there are questions, both political and cultural, regarding who is (and who is not) Sápara. As Maria, a Sápara woman with whom I worked closely for several months, explained to me, the Sápara have allowed people of other indigenous nationalities to live within their territory, and there are few *puro puro* (pure-pure) Sápara remaining.[3] As outsiders have become increasingly interested in the resources found in Sápara territory, including gas, oil, and now "environmental services", people of other indigenous nationalities have begun to claim ownership of the land, causing conflict over who has the right to claim that they are, in fact, Sápara. The political influence of non-Sápara Indigenous people over their territory is an ongoing issue. In an interview in 2013, Kléver Ruíz, President of the Sápara Nationality in Ecuador, explained that:

> ... there are several nationalities living within the territory: Achuar, Shuar, Kichwa and Sápara [...] An agreement was signed on behalf of the Sápara, without consulting with communities, between the Ministry of Hydrocarbons and Mr. Basilio Mucushigua who is of Andoa nationality, but says he is Sápara and that he represents our nationality. But he is illegally elected as a representative of the Sápara, with the support of Shiwiar and Achuar communities found within our territory.
>
> *(Tegantai, 2013)*

The Province of Pastaza, in which the Sápara territory is located, is the largest province of Ecuador. Most of its 29,800 km^2 is within the geographical region of the Amazon (Pastaza, 2010). Like most of Pastaza, Sápara territory is located in isolated rainforest with no road access. Partly for this reason, it has avoided major colonisation, industrialisation, and exploration, and has preserved much of its biodiversity (Gari, 1999, p. 23). Thus, the Sápara people are at the very early stages of market integration, which began with a small number of families planting cash crops such as cacao or peanuts (López & Sierra, 2010, p. 358). However, exploration on their land has identified at least ten significant oil deposits, which is a major concern for the Sápara people. The history of oil extraction in the country has had a devastating impact on Indigenous Peoples throughout the Amazon region and has caused some of the worst environmental damage in history (see e.g., Zelman, 2011).

Sápara territory in Ecuador is comprised of 298,000 hectares of Amazon rainforest, of which 225,000 hectares have been officially recognised with land titles. Within this territory, there are 23 communities totalling approximately 360 people (TIG, n.d.), made up of Sápara, and other Indigenous people. It was in one of these communities, at around 6 am on Friday June 13, 2014, that I awoke, in a tiny tent, hot and dehydrated, hungover, and with a lot of work to do.

The "Workshop"

I stood in front of 21 Sápara people: 14 men and 7 women. The *chicha* served with breakfast, which everyone assured me would undo some of the damage from the night before, had helped settle my head, but had the opposite effect on my stomach. It was not supposed to be like this. The room was not a room, as such, but was instead an open-sided house with a pitched roof, a fire pit in the centre, and benches around the edges. There were also logs and an upturned canoe that had been brought in as makeshift seats. In one corner was a solid wooden bed with no mattress, a table really, where the two oldest people slept. They were the parents or grandparents of most of the people in attendance, and they were apparently interested in what the white man had to say, and so some of the Indigenous men had helped run a cable from the generator near the disused schoolhouse for the projector. Some of the children were busying themselves, splashing one another from a bucket of river water, and two were playing with a small monkey that was tied to a pillar of the house, poking it with a stick. The high-pitched noises the monkey was making were distracting me from what I was meant to be doing: facilitating a workshop on gender and climate change, in Spanish, which like most of the people in the room, I spoke poorly.

When I arrived in the Sápara village – seven houses, deep in the forest where an extended family of 48 people lived – I had done so with a plan in mind. The IPCCA had given us a structure and a methodology for a workshop, which Maria would conduct while I assisted. Following that, I would also assist her in writing a report on our findings. But following our arrival, the first day was spent drinking *chicha*... for hours, and hours. *Chicha* is an alcoholic drink that is made by the women in the community. They chew raw yucca, spit it into a carved wooden trough that looks like a miniature canoe, and then it is left to ferment in large ceramic pots for several days. It is typically offered to arriving guests and is, I learned, a *very* important part of the culture. It tasted terrible to the uninitiated, and knowing it was made from someone else's saliva made it... less than appealing. As the evening progressed, the mood became more and more festive, with dancing and singing, drinking and vomiting, and it soon became apparent that the first day of the workshop would be a write-off.

But there were still two days left.

"Can we start?" I asked, already concerned about the time.
"*Sí, sí.*" said Maria.
"What do you want me to put on the projector?" I asked her.

As requested, I had taken two days to travel back to Quito from Puyo, the small city on the edge of the rainforest where I had met Maria, to borrow a projector from a friend of a friend, and to collect my laptop. Maria had said it was *absolutely* necessary for the workshop.

"I don't know, whatever you have" She responded.

I had nothing.

At that moment, a concern that had been lingering in my mind was realised. In the days before travelling to the community, I had begun to think that Maria and I had different ideas about our roles. She had given me full control of the planning, the budget, the logistics, and the shopping, and I was worried that I would be left responsible for delivering the workshop, too. So, I had asked multiple times:

"Are you sure you are comfortable conducting the workshop? You have read the methodology and understand it? Is there anything else we need?"
"*Tranquilo*. It will be fine", she reassured me each time.

But now it was clear that it would not be fine; I would be the one conducting the workshop, and I was absolutely *not* the right person for the job. By now, Maria had returned to her seat and continued chatting and drinking *chicha*.

With the projector obsolete, I set about trying to facilitate the workshop. The first task was to split the group into men and women. That part was easy enough because the women were already grouped together in one corner, while the men were in two groups on the opposite side of the house. One of the youngest women walked back and forth between them serving *chicha* from a large ceramic bowl. I began explaining the task in my best Spanish, and Maria gestured to her partner to stand up and interpret into Kichwa, but she only seemed to translate the last few words of each sentence – "men and women", "small groups", "make maps", "the village".

Everyone obliged, doing as they were asked without question, and then the men and women came together to examine the differences between the maps they had drawn. After this, they separated again to list the many different resources that are harvested from the rainforest, where they are found, and who (men or women) have the right and/or responsibility to use them. Slowly and painstakingly the tasks were worked through, followed by a general discussion about "*Buen Vivir*" and climate change, and other associated themes (including, REDD+ and oil extraction). Most of the discussions were in Kichwa, and so I learned little and spent much of the time worrying about how I could possibly write the report for the NGO.

By early afternoon, the heat and the rain (which had been fluctuating between heavy and torrential), combined with the general lack of interest among the people from the Sápara village, no to mention and dwindling enthusiasm of the newly appointed "facilitator", brought the workshop to a sluggish end. Some of the men left to hunt, and everyone else seemed happy to return to their day. I, with a belly full of *chicha* and caiman, packed away the paper, the pens, and the useless projector, feeling as though I had wasted the day. Maria said she felt that it had gone well. In the evening, a television and DVD player, which was shared by the village, were set up in the house where the workshop had been. While the community gathered around to watch *Rush Hour 2* (dubbed into Spanish), I sat in the corner near the bed of the elderly couple and drank *chicha* with them until they fell asleep. I then returned to my tent, a little drunk, very tired, and very disheartened.

The following day the rain had taken its toll. And so, the remainder of the workshop was cancelled. I was, in all honesty, relieved. I was happy to help dig trenches along the edges of the *pista*, rather than make another awkward attempt at extracting knowledge and information from the Sápara people. One of the women with whom I was digging explained that this was the real way that climate change affected them: more rain meant being more isolated, and doing more work, especially for the women. As well as being a welcome relief from my "facilitator" role, I had another reason to make sure the *minga* was a success: without it, the plane that was coming to collect us the following day could not land, and at this stage, I was more than ready to leave. It was worth the pain in my back and every one of the blisters on my hands.

The Power to Choose 1: Who Speaks, and Who Listens?

After spending only a few days in this small, isolated, Sápara community, and feeling like I had learned almost nothing, I found myself in a situation where my newly acquired "knowledge" put me in an unwarranted (and unwanted) position.

Maria's office was the headquarters of an Indigenous women's organisation and also her home when she was not in her territory, and she had invited me to work there after the trip while I was back in Puyo. So, in a building with three open sides, with dogs, chickens, and children running in and out, and the racket of rain on the tin roof, I sat trying to speak to the IPCCA on Skype. As part of my role, I had been asked to accompany Maria to a conference in Lima where the Peruvian government and other "experts" would discuss "Community Perspectives" on REDD+. But as it turned out, she was unable to attend, and now the IPCCA wanted me to choose an alternative: someone *I* believed could represent the Sápara people and who could speak on their behalf. It was a very uncomfortable position to be put in.

I asked Maria for her recommendation, and she suggested two people: both young women who were involved in the women's organisation, and neither of whom I had met before. I sat with each of them separately and asked if they wanted to go and whether they felt comfortable speaking in front of people. Both women seemed nervous, and it felt as though I was interviewing them for a job. I spoke with the NGO and said that I was not willing to make the decision for two reasons. First, I did not feel like either of the people Maria had suggested would be very comfortable speaking at a conference in front of a group of politicians and "experts", but more than this I felt that I should not be deciding who is (and who is not) a legitimate representative of a community I knew almost nothing about. "Well, somebody has to go, it is in the Terms of Reference", was the response.

Feeling frustrated, I suggested that Maria and the two other women decide among themselves who should go. At first, I felt good about this, and I thought I had made the right decision by "allowing" the Sápara people to choose their own representative (though it was by no means a democratic or participatory process). Maybe I *was* the right person for the job, I thought; I was, after all, there to help. But over the next few days, it played on my mind. The decision not to make the decision was still, in the end, *my* choice, which was made based on my existing ideology: i.e., that Indigenous Peoples should be able to choose their representatives through a process of self-selection. And yet by virtue of my position (as a white British man who happened to have a couple of degrees, but no knowledge of the Sápara community), I was able to give an entire group of people permission to do something that they really should not need my permission to do. I also wondered if I would have given up my power so willingly had one of the suggested people been "unsuitable" in my eyes. I am not sure I would have.

*

In a five-star hotel in Lima, there I sat, a white man on a panel with three other people: a Black woman from a University in the United States, a mestizo man from a University in Mexico, and Luisa, a Sápara woman. I wondered if it had been designed that way. We were there to discuss "Alternatives to the REDD+ Vision: Non-carbon Benefits, Community Perspectives, and Possibilities". A few hours before, the head of the IPCCA had suggested that my talk be changed from a critique of REDD+ and Socio Bosque, which reiterated some of the things I had learned in the Sápara community, to a general overview of the project and what co-benefits it offers the community, and also asked that I reduce it from 20 minutes to 10. I was glad to speak less but frustrated with the willingness of the NGO to conform to the status quo. Nonetheless, I said I would try to include some potential "co-benefits", but that it was not the "community perspective" I had heard. I was introduced by the convenor of the panel:

> George Byrne has a degree in Latin American development, a master's degree in international relations, and a master's degree in social research methods, his PhD research is on REDD+ and Indigenous Peoples.

I remained seated so that I could read from my shoddily abridged notes:

> To begin, it is worth considering how serious we are about considering community perspectives on REDD+. When I look around the room, I see very few people *from* the communities that will be most affected by the projects that are being designed or implemented under the banner of the programme.

The head of the NGO looked uncomfortable, and I felt self-conscious about giving my first ever presentation in Spanish.[4]

> From the beginning, both Socio Bosque and REDD in Ecuador have been criticized and resisted by environmental organizations and indigenous groups. For example, in 2009 and 2011, respectively, CONFENIAE and CONAIE rejected negotiations on Socio Bosque, arguing that it would privatize indigenous land and take away their rights to manage their own resources.
>
> These are real concerns, and the MAE is aware of the shortcomings of REDD+ regarding indigenous participation. The results of a 2013 report by the Organization of Social and Environmental Standards on REDD found that:
>
> *...the communities visited in Ecuador did not participate in the process of consultation or free, prior and informed consent. The decision to join Socio Bosque was communicated in the general assemblies, which exclude marginalized groups, such as women, youth and the elderly.*

I continued, shifting my focus to the shortcomings of my own recent workshop, avoiding eye contact with the head of the NGO:

> To ask people to sit in groups and discuss "what *buen vivir* means for you", to make maps of their uses of the forest, or to ask how climate change has affected the community does not produce the expected results. [...] The problem is in the way we frame questions about REDD+ and the methods we use to get answers.
>
> This is characteristic of the existing approach to studying the impact of REDD+ and similar projects. We ask complex questions, all of which are formulated within the language and ideology of "sustainable development". But we want simple answers.
>
> It is expected that indigenous groups, with highly diverse forms of knowledge and epistemologies, cultural practices, and social structures will provide information that fits the predefined categories of responses.
>
> On the other hand, it is hoped that this can be achieved on a time scale and to a budget that conforms to what governments, NGOs, and donors consider acceptable.
>
> The way in which the co-benefits and impacts of REDD+ are analysed and communicated with indigenous groups does the opposite of what they claim: they do not include Indigenous Peoples, they continue a process of colonisation of indigenous territory. This is what I was told by the Sápara people, but I do not want to speak on behalf of them. So please, listen carefully to what my friend from the Sápara community has to say.

I did not receive the formality of a quiet applause that the previous two speakers had. But I was fine with that. They had spoken about the inevitability of some form of a PES model and how, despite its problems, REDD+ was moving forward. They said that it was still in the process of formation, and so now is the moment of opportunity to influence how it works and make co-benefits as equitable as possible. COP20, which would be held here in Lima in November, would be the time and place to make sure this happened. I felt my face become red, and I hoped that whatever Luisa said would make me feel like less of an idiot. Her contribution was brief, and she spoke in Spanish, quickly and passionately. I found it difficult to keep up, still wondering whether my Spanish had been clear and worrying about what the NGO thought. And so, my notes were brief, but I was relieved to hear what she said. The convenor introduced her:

> Luisa, from the Sápara people in Ecuador.

She stood and began:

> My name is Luisa. I am a woman from the Sápara nation in Ecuador and a representative of our women's organisation [...] We, the Sápara people, live in a beautiful and rich land where we have everything we need for our survival [...] They tell us that we are poor, but our culture is rich, and we have knowledge of the forest that nobody else has. But now our land and our lives are under threat from contamination and oil [...] We were told that Socio Bosque would protect us from *los petroleros* and that we can benefit from REDD, but it is the same, it is a lie.
>
> *(L. 2014)*

Luisa's presentation echoed what I had been told in the Sápara Community:

> For me, the good life is to live in the best way with my family and my community. To unify our family in a dialogue of peace with the community and nature. Living better, sharing, working and being united. But we are afraid of oil and government. They want our land. Money cannot buy it from us. No amount would be sufficient because money is temporary and after it is gone, we would have nothing. Our land is permanent.
>
> *(A. 2014)*

> We must keep the oil below the ground because if we don't, we will have contamination, and if this happens nobody will be able to live here, we won't survive. This is how the economy works; it is unequal. It is the same in the city. When indigenous people go there, they don't get rich, they stay poor. They say we are poor now, but we have a great territory, and the land is all the money we need.
>
> *(M. 2014)*

Luisa took her seat and received a polite applause. But all of the questions that followed were directed at the three other panellists:

> How much money do projects in Ecuador get per hectare of rainforest?
> How do you see the future of carbon markets developing?
> How is the carbon in rainforests measured, and does it include subsoil carbon?
> Are there any examples of communities where REDD+ projects have been effective and inclusive?

In the afternoon, Luisa left to visit the market and buy gifts for her family while the rest of the room returned to discussions about what a "good" REDD+ strategy might be in Peru and how "co-benefits" might be maximised. The title of the day had been misleading. Though everyone had their criticisms and concerns about REDD+, the conference was not about finding "alternatives" or hearing "community perspectives", it was about finding ways for communities to work within it.

Recognising the Colonist Within

I came away from my first trip to Sápara territory and the conference in Lima feeling as though I was more a part of the problem than the solution. I became more aware of how colonialist thinking had seeped into my research methods and methodology. In academia, it is rarely acknowledged, at least outside of decolonial and post-colonial scholarship, that colonialism is an ongoing process and that we as researchers are a significant part of it. Even the idea of post-colonialism raises an important question, as Bobbi Sykes, an Australian Aborigine activist, asks "What? Post-colonialism? Have they left?" (Smith, 1999, p. 24). Well, no, we have not. It is uncomfortable to admit that as a researcher I have valued the knowledge I already have, and the research I am trying to produce, more highly than the people I "study" and the knowledge that they have. Even when adopting a "participatory approach", there is a risk that extracting knowledge benefits people elsewhere and leaves communities at best unchanged, or even worse off than they were before (Wilmsen, 2008, p. 135). From the perspective of some Indigenous people, governments and NGOs have often done the same, attempting to extract the information they want using methods that give *their* ways of knowing supremacy, Rarely is it asked what or *who* this type of research, and the immense labour of information retrieval, is really for (Beverley, 1999, p. 38). Some months later while at COP20 during a "Public Hearing on Indigenous Peoples" Concerns and Perspectives', a group of Indigenous people performed a short play that was intended to illustrate a typical example of an interaction between their community and two outsiders. The characters were an Indigenous "Wise Man", his daughter, a Public Servant, and a "researcher" that they named "Mister Doctor Professor Koch". The whole performance was very funny and cleverly communicated a number of ideas that are related to this book, but one particular part of the interaction is pertinent here. After gradually persuading Mister Doctor Professor Koch that Indigenous Peoples have knowledge that the West does not, the dialogue continued as follows:

> Wise Man: Your scientists must listen more closely to us. The answers come from nature itself. You have disturbed nature and you will have to ask pardon and reconstruct the relationship with the forces of nature. You must learn to live with respect for all that lives and that has a spirit.

You have to *learn to listen* to us as original peoples, with traditional knowledge about the balance of nature.

You see us as a *problem*, but we are, in fact, the solution. Your sciences are good, but they are not sufficient because you do not understand nature like we do. We need to talk with one another more. With respect and equality. Only if we work together can we solve this problem.

Mr Doctor Professor Koch: Very well, Mister. When will you be able to come to Washington to speak with our panel of professors? We will take all of your knowledge, and afterwards, you can go back

Wise Man: No sir; you still don't understand. *It isn't "my" knowledge*; it is a knowledge that we have constructed over centuries; through collaboration and exchange between many individuals and communities. This knowledge cannot be seen as separate from the way in which you view the world.

If you don't know how to use this knowledge in the correct manner, you won't be able to use it at all. The use of traditional knowledges goes hand in hand with respect for nature and for the supernatural; it is not like your laboratory knowledge.

On my part, I arrived in the Sápara village with set questions, set methods, and a set timescale within which I was expected to extract "useful" information or knowledge on behalf of the NGO I was working for. The IPCCA described itself as having "emerged as an innovative response, bringing together indigenous knowledge and science in a process which links biocultural realities with complex global processes". It aims to use "biocultural methods and tools [to] involve communities from around the world in the assessment of climate change and local well-being *and the development of evidence-based responses* [emphasis added] for climate change adaptation" (IPCCA, 2013).

It is no coincidence that my self-critique through the lens of decolonial theory began to surface during this time. Though I had been very critical of the idea of "development", and particularly "sustainable development", in the past, and felt I had a good understanding of how environmental projects and programmes might facilitate hegemony, my focus had begun to shift toward a reflection on my own role in this process. I had been reading *Decolonizing Methodologies* by Linda Tuhiwai Smith (1999) over a few weeks, and after returning to Quito from Peru, I read an article by Juanita Sundberg called *Decolonizing Post-Humanist Geographies* (Sundberg, 2014, pp. 33–47), which articulated many of my emerging concerns about my position. One phrase in the article that hit home for me was "learning about the colonizer who lurks within" (Regan, 2010, p. 11, cited in Sundberg, 2014, p. 39). This is an unnerving prospect, but I agree that it should be a fundamental part of conducting research in indigenous communities or with Indigenous Peoples elsewhere (or with any marginalised people). It requires the researcher to reflect critically on one's own ontological and epistemological assumptions, relinquish control, and to "unlearn" much of what we have been taught. A key part of this is what Johannes Fabian calls "going out of our minds [...] leaving one's comfortable psychological, political, and discursive place" (Fabian, 2002, cited in Sundberg, 2014, p. 40). The method of doing so is described by Spivak as "homework", the process of reflecting on one's own position and privilege in society, particularly the "privilege of sanctioned ignorance that allows the perpetuation of silence about ongoing colonial violence" (Spivak, 1990, cited in Sundberg, 2014, p. 39).

I decided that, when I returned to the Sápara village, I would do so without set questions, without the NGO's methodology, and without a specific plan. I would spend *time* with people, speak with them, and *listen* to them. I would try to put aside my frustration with the time wasting and disorganisation, and I would set aside the feeling of not being "productive". I would listen to stories without questioning how true they were, but instead trying to understand what they meant to the people who were telling them. But this, I found, was harder than it seemed.

Going Out of My Mind?

It was a month to the day after the conference in Lima that an opportunity arose to return. There were two planes flying to the Sápara community and on-board would be a photographer from a US NGO, an independent Ecuadorian photographer, and two members of an Ecuadorian NGO. Following the painfully slow progress of the report for IPCCA, Maria had been asked to organise another trip for a second round of data collection. I was again invited along. And so, with a head full of "decolonised" approaches and methodologies, I set off once again from Quito to Puyo, met with the rest of the visitors at the tiny airport in Shell Mera, and took the 45-minute flight into Sápara territory.

By the time the plane took off, I was already struggling to not get annoyed. After being told to arrive at 9 am, I waited at the airport. Getting impatient and worried that I had the wrong day, I finally called Maria at 10.30 am to find that she had not yet left her office in Puyo. At that point I had no choice but to sit on my backpack and wait until everyone else arrived. We finally boarded the plane at midday. Once we arrived in the community I began to relax, reminding myself of what I had read about the importance of "relinquishing control" as Fabian put it (cited in Sundberg, 2014, p. 40). And I think I did manage this, to a large extent. I stopped asking questions and *let* the Indigenous people speak. I observed the interactions between the NGO staff, the photographers, and the community. But I found it difficult, and I was struck by how easy it seemed to be for the other visitors. I wondered again if I was the right person for this type of thing. I tried not to judge the Indigenous people, or to interfere, or to impose my values or ways of being, but I couldn't help coming to the rescue when I saw a child dangling a kitten over a fire. Everyone has limits.

The first workshop was conducted with the women of the community, led by one of the photographers, Felipe; it focussed on oil and gender and applied an arts-based approach. The women who attended appeared to enjoy it. They took pieces of paper with a rectangle drawn in the middle, decorated them around the edges, and wrote a comment about climate change at the bottom. These would later be used as frames for their portraits, which were taken by Felipe. One comment read:

> My name is Ana Santi, Sumak Kawsay. I want for my children to be able to live well. If the oil company comes in, we are not going to live well. I am raising my voice as a Sápara woman for our children and those to come. I was born on the banks of this river. That's why I refuse for the oil company to occupy our territory.
>
> *(Santi, 2014, cited in Jacome, 2015)*

The rest of the workshops were more conversational and focussed on oil contamination and the recent exploration in nearby concession blocks. The two women from the Ecuadorian NGO hung maps of the region on a wall, and they talked about the risks of contamination and a little about Socio Bosque while one of the photographers filmed from a corner and the other took photos. I took some photos too, but I reminded myself that it was not why I was there. I tried to listen.

> I have just travelled back from my community near Block 10. My grandfather and uncle worked for the oil companies and now they have both died from cancer. The same thing will happen here. We need a workshop in my community, too, and the maps so we can see where the pumps are. Most of the people are against oil, but some still want to work for them to earn money.

The Indigenous woman was visibly distressed and struggling to speak. I took a note: "she is visibly distressed and struggling to speak". Another woman stood and said:

> Oil and Socio Bosque are the same thing. I know. I spoke with the head of Socio Bosque and he told me that they will pay indigenous people,

> but then they will still be able to exploit the oil on the land. And they won't be able to do anything about it because they have already been paid. It is just a new colonialism.

There was a long discussion about whether the person who signed the Socio Bosque agreement had, in fact, done so. One man said it have never been signed, "it's a lie!", and another argued the signature had been forged. A woman who seemed to be quite influential stood and insisted that it had been signed, but never should have been, and the discussion ended there.

Rather than conducting interviews, or a workshop, or taking lots of pictures, I spent time with people. I listened to what they had to say, though they rarely said much to me. I handed my camera to the children and let them take pictures, holding my breath every time they almost dropped it. I weaved baskets with men and went with a group of women and children to make *chicha*. I drank it and pretended to like it, and even began to get used to it. One night, a boy of around ten years old asked if I wanted to go frog hunting. I did not, but I did it anyway. I watched a woman making ceramic bowls, and she asked if I wanted to buy some. I did not, but bought some anyway.

After one of the workshops, I sat with a man weaving a basket. It was a man's job, I was told. He did not speak much, but he did say:

> You know, Jorge, *buen vivir* is to live here with my brothers and my family, as part of a community. Now we go away to work, away from our family, to earn money for months or a year [...] we only want to live without contamination and without chemicals. We buy chicken and other food from outside sometimes, and even that has chemicals in it. That is why I don't want to live in the city. We come back from outside with cancer and other illnesses. We just want our people and our culture to survive. My two sons are in Puyo and Sarayaku territory studying and learning how to defend our territory.

I, of course, wrote this down. "Good quote!", I thought.

On the morning of my last day in the community, the television and DVD player were set up once again, but it was not a Hollywood action movie, this time. The film was *Crude*, a documentary about the lawsuit against Chevron. I had seen it multiple times before, so I went to the back of the room to write some more notes:

> "What are you writing?" a young Indigenous woman asked.

I explained that I was writing about the workshop and the community.

"Why do you always write so much?"
"So I don't forget"
"You will remember what is important", the young woman said.

She sat and offered to explain how everyone was related. We drew a family tree, and we wrote each other translations of words in my notebook. I translated from English to Spanish, she translated from Spanish to Kichwa.

"You don't want to watch the film?" I asked.
"No, we watch it all the time!" she said, and she rolled her eyes.

The Power to Choose 2: Money… and Other Misunderstandings

When I returned to Puyo. I felt quite good about the experience I had on my second visit to Sápara territory, and I thought I had begun to build some relationships with the people I had met. Though I did not come away with any answers, I felt I understood the questions I had been asking a bit better. I was also sure that subsequent visits would allow me to continue to develop my approach to working with the Sápara people and possibly even have a positive impact on them in some way. But as it turned out, I would never return to Sápara territory.

In the build-up to the second visit, it had become increasingly clear that Maria and I had very different ideas regarding our roles and responsibilities. As I understood it, I was her assistant, and she would be conducting the research. We were both being paid by the IPCCA, though neither of us knew how much the other was getting. At first, I had offered to work on a voluntary basis, but as the NGO added more and more tasks to the Terms of Reference, I decided to accept payment and use it to pay for flights for further trips to the Amazon. It seemed like a sensible way to help fund the remainder of my research, but it became an unexpected problem that ultimately led to a breakdown of my relationship with Maria and, as she was the gatekeeper, this meant I would no longer have access to the Sápara community.

Money had been a point of contention from the very beginning of our time working together. On the day we met, I had been given the task of developing a budget for the visit to the Sápara village, which involved discussing the expenses with Maria. As we went through the list of possible requirements, I asked which were necessary. Buy a camera? Yes. Pay the community for their time? Yes. Food, flights, paper, pens, water, fuel for generators? Yes. Hire a projector? No, buy one. Every possible item was required, and each should be given the highest possible allowance. I sent the budget to the IPCCA feeling a little apprehensive, but to my surprise they authorised everything but the projector. Once in the community, it seemed to me as though Maria and the other Indigenous people had little interest in the workshops, but they went through the motions without complaint. Many of the items we brought with us were never used, but there were further requests to come. For example, the President of the community asked me to return one day with medical supplies or set up a health fund for them, another man asked for my tent, and Maria asked me to leave the camera and projector in the Sápara village (I did leave the camera, along with everything else we had bought, but the projector was borrowed and so had to be returned, and I needed my tent).

This pattern continued throughout the process of attempting to write the report. When I had meetings with Maria in her office or in the town of Puyo, I felt an obligation to provide food and drinks for whoever attended, which was sometimes just us, but at other times would include her partner, her child, and various members of her extended family. Sometimes I did not even know who it was for. I think that this was partly my fault because at our first meeting, I offered to buy lunch, which was intended to be a gesture to thank Maria for her time. But the second time, when we went to buy food for us and her family, we reached the register and Maria just walked out of the shop, leaving me to pay. Then she began to just ask for money: for phone credit so she could call me, for transport, for food, and various other small things. None of this was included in the budget, but I paid for it anyway. I was glad I had agreed to be paid.

A situation arose that gave me some insight into what might be going on. On the way back from the conference in Peru, I checked with Luisa how much cash she had left because my credit card hadn't been working at the airport. As she had requested, I contacted the NGO before we left and asked for extra money for childcare and for a day to visit her family on her way back. But Luisa told me she had spent all of it on gifts from the market, including the money for our transport from the airport in Quito. I tried not to get annoyed, but I was, and I explained to her that we now had no way of getting back to the city. I asked her what she thought I should do "I don't know... *tranquilo*. It's fine! You will find some money", she replied, and then she went to sleep. As frustrating as it was to feel as though Luisa did not care about how or where I would find the money, she was, in fact, correct. I did manage to get cash when we arrived at the airport in Quito.

I began to think about how the people I was working with viewed me, and why they were working with me at all. I had started to realise that the relationships I had hoped would be equal, and might even lead to friendships, were something else. They had become interactions – or transactions – limited and defined by preconceptions we all held about each other, by an unequal balance of power and resources, and by having different things that we wanted to achieve. To me, *they* were the Indigenous people who I was supposedly here to help, and *I* was a researcher with good intentions. But I had, and still have, little insight into how they viewed themselves or me, or who they thought was helping whom.

In the report for the funder, I was asked to comment on the extent of "market incursion" on the Sápara community. I wrote that:

> Due to its relative isolation, the Sápara village has so far been integrated very minimally into market systems. The small amount of trade that they do have with the outside is to buy supplies with the money earned from occasional wage labour in cities and by selling a very small amount of artisanal goods to resellers in Puyo.

But later I realised that I had overlooked the extent to which the *development* industry – specifically those NGOs that seek to support the Sápara people in securing their cultural survival and in protecting their territory from exploitation – had inadvertently increased the dependence on the community on funds from outside, especially since their culture and language had been recognised by UNESCO as a Masterpiece of the Oral and Intangible Heritage of Humankind. My project with Maria was also, of course, part of this process. Moreover, it was no longer simply influencing the way that the Sápara people interact with outsiders, but had, to some degree, been internalised by the representatives of the community who work on the cultural frontier. Despite the many inequalities and injustices they had to navigate, and whether or not they shared the same ideology as their funders and supporters, the representatives of the Sápara people had become well-versed in how to make these interactions work for them.

Conclusion

The eventual end to my work with the Sápara people happened gradually between August and November 2014. I had contacted Maria many times regarding the report we were supposed to be writing together, but it had become increasingly difficult to reach her. Sometimes weeks would pass, and the NGO would contact me again asking for the draft. I would apologise and explain the situation, and they would give us another deadline. I later learned that Maria had gone "*dentro*" (inside the forest). She was in hiding in her territory because a warrant had been issued for her arrest under allegations of terrorism[5]. But with the limited data we had collected and with no contact from Maria, I was not comfortable submitting what I had written. I had also come to realise that Maria's written Spanish was poor and that she would never have been able to provide what she had agreed to. And so, in the end, I did the best I could with what knowledge I had and submitted the report, meaning we would both be paid.

Unfortunately, a clerical error on the part of the NGO resulted in my fee being credited to one of the accounts Maria controlled rather than directly to me. The NGO told me I would have to travel to Puyo and ask her to withdraw it for me, which would also mean travelling a full day by bus back to Quito with the cash. I did not feel comfortable at all doing either of these things, but I eventually managed to reach Maria via email. She had returned to Puyo from her territory and invited me to meet her there. We had dinner with her partner and another friend. I paid (I expected to by this point), we discussed the issue we had had regarding the mistake by the NGO, but she did not want to talk about why she had disappeared for so long. The next day she told me the money was not yet in her account. She asked me to wait, so I stayed in Puyo for a few days, but again, Maria went silent, and I returned to Quito. After many emails back and forth with the NGO, who insisted the transfer had cleared, I decided to give up and just try to move forward with my own research.

Then, on November 14, 2014, I received a short video message via Skype. Maria was pacing behind her partner, who was sitting at the computer:

> "What is his name again?", Maria whispered.
> "Jorge", Her partner responded.

Then Maria said:

> You know, Jorge, you must not communicate very much with the IP-CCA, at all. You have to see how it was voluntary. And I don't… I don't want to speak. Leave it! I am going to go now.

Then, at the end of the message, Maria walked away from the computer and said to her partner:

> He comes here and we let him work with us and he just wants to take money and take money.

This came as a shock to me, as when we had met for a meal in Puyo, things seemed fine. I felt I had done everything possible to deal with a difficult situation, but Maria clearly did not feel the same. I decided to do as she had requested and leave her alone, let the money go, and move on. We never spoke again, though our paths crossed a few times over the coming year.

Over a year later, I was in Paris at COP21 with an acquaintance, Elena, who had over 20 years' experience working with many different Indigenous people in Ecuador, and she had heard *chisme de la selva* (jungle gossip) that something had happened between me and Maria. She asked, and so I explained the situation and found myself making excuses for Maria. Elena told me that I was being patronising and that I should have pushed it, because if I did not it would just happen again. I don't know which course of action was right, but either way, Maria had made her position clear: she did not trust me, and she did not want to work with me. And as she was a highly influential figure among the Sápara, and the gatekeeper to the rest of the community, working with them would have been impossible without her. At a loss, I eventually decided to go to a place where I knew Indigenous Peoples, not only from Ecuador but from all over the world, would be: the Indigenous Pavilion in Lima. This is where I would encounter COICA for the first time and hear of REDD+ Indígena. But as the chapter that follows illustrates, the experiences that led to this "choice" had a profound effect on my research, and on me.

Notes

1 Michell (2012) used the metaphor to describe a "quest for knowledge".

2 As with the ambiguity surrounding the population of the Sápara, estimates of how many people still speak the language vary considerably. Approximately six elders (all over 70 years of age) are believed to speak Sápara fluently in one of the two remaining dialects (UNESCO, 2014). The majority of Sápara people now speak Kichwa, and many also speak Spanish. This is partly due to Kichwa developing as a lingua franca among Indigenous Peoples of the region in the first half of the 20th century, and partly the result of Sápara people (mostly men) travelling to work for oil companies and various plantations around the city of Puyo, where Spanish is spoken extensively, in the second half of the century (Ushigua & Viatori, 2007, p. 9). The loss of language has been a significant obstacle in proving to outsiders, including the Government of Ecuador, that the Sápara people are a distinct indigenous group with a unique culture and ancestral claims to their land. Thus, contemporary attempts to revive the language are both culturally and politically important.

3 "Pure pure" is a translation of the locally used Spanish description "*puro-puro*". Another factor that makes identifying how many Sápara survive is that it is common for Sápara people to marry, and have children with, people of other indigenous nationalities that live in the region. There is some debate about whether children with one Sápara parent are indeed Sápara. People with two Sápara parents or "Sápara puros" are far less numerous. Those who are considered "puro" have usually learned the Sápara language and have

maintained some degree of fluency after learning Amazonian Kichwa or Spanish. Those who identify as Sápara usually assert that "raway Sápara" (Sápara blood), passed on from parents, carries Sápara characteristics and so is integral to being Sápara (Viatori, 2005, p. 81). A number of people that I spoke with regarding this explained to me that, today, children with a Sápara father are considered Sápara, though not "puro", while children with a Sápara mother and a father of a different indigenous group are not. This patrilineal construction of Sápara identity has not always existed, rather it is a result of attempting to define who can and cannot speak on behalf of the Sápara people. Part of the reason for this is that as marriage between indigenous groups has become more common, it has been customary for women to move to the community of their husbands. Thus, children with Sápara fathers will generally grow up in Sápara communities and will identify with the culture. The second reason is that it has become increasingly important to define who is Sápara, as this is a crucial part of being recognized as a distinct indigenous culture and gaining legal rights to ancestral land.

4 In a conversation at dinner, I mentioned I had been nervous about this, and someone who worked in the Peruvian Government said, "you should have just spoken in English, everyone in the room would have understood". I responded to this by saying, "Luisa wouldn't have, and it was her community that I was talking about".

5 These allegations were made by the government of Ecuador and were related to Maria attending protest against the oil extraction and campaigning against Socio Bosque. However, I will defer discussing "terrorism" until the next chapter, where it can be dealt with in the context of fear.

References

Beverley, J. (1999). *Subalternity and representation: Arguments in cultural theory.* London: Duke University Press.

Curwood, S. (2004, Jul 30). Can this tribe be saved? Living on Earth. Retrieved from https://www.loe.org/shows/segments.html?programID=04-P13-00031&segmentID=5

Gari, J. A. (1999). Biodiversity conservation and use: Local and global considerations. Science, Technology and Development, Discussion Paper (7), Cambridge, MA: Center for International Development and Belfer Center for Science and International Affairs; Harvard University.

IPCCA (2013). *Welcome to IPCCA.*

Jacome, F. (2015).Third Eye. Retrieved from http://www.thethird-eye.co.uk/guardians-of-life-amazon-women-ecuador-by-felipe-jacome/

López, S., & Sierra, R. (2010). Agricultural change in the Pastaza River Basin: A spatially explicit model of native Amazonian cultivation. *Applied Geography, 30*(3), 355–369.

Michell, H. (2012). *The Canoe Trip: A Northern Cree Metaphor for Conducting Research.* Retrieved from http://journals.uregina.ca/index.php/ineducation/article/view/4/489

Pallares, C. A. (2001). *Kwatupama Sapara, Palabra Zápara.* Abya-Yala: Puyo.

Pastaza (2010). *Ecology of Pastaza, Ecuador.* Retrieved from the http://www.pastaza.com/en/ecology/

Smith, L. M. (1978). An evolving logic of participant observation, educational ethnography, and other case studies. *Review of Research in Education*, 6, 316–377.

Smith, L. T. (1999). *Decolonizing methodologies: Research and indigenous peoples*. London: Zed Books.

Spivak, G. C. (1990). *The post-colonial critic: Interviews, strategies, dialogues*. London: Routledge.

Sundberg, J. (2014). Decolonizing posthumanist geographies. *Cultural Geographies*, *21*(1), 33–47.

Tegantai (2013). *Entrevista: Kléver Ruiz. Presidente Nacionalidad Sápara*. Retrieved from http://www.agenciaecologista.info/petroleo/457-entrevista-klever-ruiz-presidente-nacionalidad-sapara

TIG (n.d.). *Zápara*. Retrieved from http://www.territorioindigenaygobernanza.com/necu_08.html

UNESCO (2008). *Oral heritage and cultural manifestations of the Zápara people*. Retrieved from http://www.unesco.org/culture/ich/index.php?RL=00007

UNESCO (2014). *Oral heritage and cultural manifestations of the Zápara people*. Retrieved from https://ich.unesco.org/en/RL/oral-heritage-and-cultural-manifestations-of-the-zapara-people-00007

Ushigua, G., & Viatori, M. S. (2007). Speaking sovereignty: Indigenous languages and self-determination. *Wicazo Sa Review*, 22(2), 7–21.

Viatori, M. (2005). New bodies, ancient blood: "Purity" and the construction of Zápara identity in the Ecuadorian Amazon. *Journal of the Society for the Anthropology of Lowland South America*, *3*(2), 175–195.

Wilmsen, C. (2008). Extraction, empowerment and relationships in the practice of participatory research. In B. Boog, J. Preece, J. Zeelen, & M. Slagter (Eds.), *Towards quality improvement in action research: Developing ethics and standards* (pp. 135–156). Rotterdam: Sense Publishers.

Zelman, J. (2011). *Chevron Accused of 'World's Worst Oil-Related Disaster' in Ecuador: Alleged Evidence Submitted in Lawsuit*. Retrieved from http://www.huffingtonpost.com/2011/01/24/chevron-accused-by-ecuado_n_813117.html#s228930

6

FEAR, THE REAL, AND THE OTHER

Introduction

All other sounds were rendered inaudible by the heavy drone of the propeller, which was just about visible through the haze of mist on the windscreen. After hours of waiting in the tiny airport at Shell-Mera,[1] a small Amazon frontier town in the foothills of the Andes, we were finally on the runway. I was about to make what felt like the last leg of a journey that had begun during my PhD application almost two years before when I had decided that I would travel to the Ecuadorian Amazon in a bid to find out how REDD+ would affect those people who live in, and whose livelihoods depend upon, the rainforest.

In that moment, I felt calm for the first time in months. The sound that shrouded the aircraft was almost indistinguishable from silence; a comforting white noise that masked even the passing of my thoughts. The pressure that I had felt building over the past few months, the worry that I would not be able to complete my fieldwork at all, had begun to fade. I was both relieved and excited to finally be aboard the plane. As we accelerated, our backs were pushed against the seats and the fine mist in the air condensed into droplets on the window: it was about to rain. Shell-Mera is less than 10 kilometres away from Puyo. The word "puyo" means "cloudy" in Kichwa, and it is a fitting name for a place where the humidity, haziness, and greyness give the impression that it could rain at any moment. It always seemed to be raining or about to rain in these forest towns. Puyo was where I had met Maria in person for the first time a few days before, and we had spent some time together in her office in a barrio

DOI: 10.4324/9781003341864-10

on the periphery of the town preparing for our trip. As the panels and panes rattled and creaked, water slowly began to creep through the broken seals, and the six-seat Cessna 206 strained for all it was worth to reach the necessary speed for take-off.

Before we had boarded, everything (the pilot, the five passengers, the cargo, and our luggage) had been weighed and accounted for, and because we were apparently at the absolute limit, we had been asked to leave a few non-essentials behind. I was a little worried that they had forgotten to weigh Maria's child, who had been playing at the side of the runway while we loaded the plane. But they were only small, and I reassured myself that the pilot would have noticed if it *really* mattered.

I stared out of the window, watching and waiting like I always do during take-off. As the world speeds up, I look down at the ground, waiting for it to be pulled away, or I watch the buildings at different distances cross over each other. The parallax makes it seem cinematic; the window becomes a lens. I like flying: it doesn't feel real. The wheels left the runway and the vibrations subsided, and it was then that I noticed Maria's right hand on my left wrist, holding it tightly. I did not know how long it had been there, and I had only noticed it at all because of the sudden squeeze that had coincided with the instant we took flight: that moment where your stomach disappears for a second.

I was surprised to feel Maria's hand on mine for two reasons. Until this point, she had only ever shaken my hand once, very briefly, when we first met at my hostel in Puyo, and she had averted her gaze when she did so. She had not seemed comfortable with me at all, which had made me feel nervous and awkward. But it was particularly unexpected because before the flight there had been some tension between us, she had seemed cold and irritable, which I had put down to the stress of organising the trip. But as I turned my attention away from the window, from the view that was now only green and brown and grey (trees, and rivers, and clouds), I saw Maria's face, and her eyes were fixed upon mine for the first time. Her posture was upright and rigid, and she seemed to be holding her breath. I immediately recognised how afraid she was, and I realised that what had seemed like impatience or irritation before the flight might have instead been her apprehension about flying.

Seeing Fear

Despite the trouble we had had in communicating before the trip, and although I was not afraid myself, at that moment I felt that the fear that I saw on Maria's face needed no real translation. I could understand it, and yet I could dismiss it because, although I felt sympathy towards her, the assumptions I then made about her fear, about *her*, and subsequently about *all* Indigenous people were shaped by my tendency to rationalise and be objective, and to minimise the experience of Others.

I only visited Maria's community in the rainforest twice, each time flying in and out on light aircraft. Four flights in total, on four different planes, each overladen with people or cargo: food, water, and diesel for the generator on the way in, artisanal goods and *chicha*, or an extra person squeezed in on the way out. I had been told stories of how dangerous these little planes could be, but I knew that they were exaggerated. Flying is the safest form of transport. On the first trip I attempted to conduct the workshop discussed in the previous chapter, and the weather was so bad that we had to wait for it to pass before the plane could come to collect us. We spent the time digging trenches along the *pista*, which had become waterlogged, and therefore useless as a runway, due to the uncharacteristically heavy rainfall. When we finally left, the turbulence did not bother me, and though we seemed unfeasibly close to the treetops during take-off, I was just glad to be leaving. Weeks later, on the second flight back to Shell-Mera from the Sápara village, I watched as the pilot drank beer from the cool-box wedged in alongside his seat, and I thought little of it, but Maria's unease made me feel uncomfortable, and a little irritated, once again. I felt as though her irrational fear of flying was being projected onto me and making *me* feel unsafe as if it were contagious. The pilot had made too many flights that day, he told us, and this would have to be his last. We were lucky to be collected at all, he said. He might have been a little drunk, and he looked very tired, but we landed safely.

On each flight, when the wheels left the ground, when we hit turbulence, or when we landed heavily on the tarmac or on the mud, Maria would grab my hand or my arm, and sometimes she would close her eyes. I would smile and reassure her with patronising sympathy, "*Tranquilo*, it will be fine". But every time I saw the same fear, and I always thought some variation of the same thing:

She doesn't even understand flying.
I shouldn't have such high expectations of her.
She is just doing her best.
It must be so difficult to live in a world
you are so afraid of and don't understand.
I am glad I am not so irrational.
It must be terrible to have such an emotional
reaction to something so normal.

I remained oblivious to how these thoughts were shaped by my assumptions and biases until far later when I began to experience fear differently for myself. Of course, before I began my fieldwork, I knew of (and had been critical of) the stereotypes and caricatures associated with Indigenous Peoples, and I had convinced myself that I could see past them to what is real. But I suppose that is the nature of implicit biases[2]; they reside in a part of the mind that seems to operate independently of what we *know* we think.

I saw fear in the Sápara village, too. In the workshops where climate change and oil were being discussed people recounted stories of sickness and death, and of contamination. And one night after too much *chicha*, when the dancing and laughing had subsided, the President of the community put his arm around my shoulder and began to cry:

> Jorge, when we get sick, we can treat ourselves with our own medicines, but now the plants we need are difficult to find, and the sicknesses we get from outside are too strong, we don't know how to treat them. We need drugs now. Can you help us get medicines? You could start a "fund"!

I listened, I heard him, and I saw his fear, and I wrote down what he said. But I did not start a fund, and I resented the assumption that I could.

On a different night, Maria told me that everyone, drinks too much, especially the men, and when they do, they sit up late, afraid of the "men in the forest". I did not understand what she meant, and when I asked her, she changed the subject, but later her partner explained to me that people who they did not know had been appearing in their territory. They said everyone wanted their land: the government, *los cólonos*,[3] and *los petroleros*, and it was not safe in the forest for them anymore. That was why men always hunted in groups, now, and why the women no longer went to the *chakras*[4] alone. But the details were always vague, and the stories were always changing. The lack of consistency made me question the truth of what they said and I took notes about my doubts:

> I am finding it difficult to know what is true and what is not. Nobody explains what they mean and the story changes from moment to moment, and from person to person. I think they are really afraid, but I don't know what of, I think they are also a bit paranoid.

These doubts about whether what I was being told was factually accurate were not unfounded. Halfway through the second trip I had realised that the community I was in was not, in fact, the community I had been told was visiting at all. It was another Sápara village down river from the one named by the IPCCA in our Terms of Reference. Nobody had mentioned this on the first trip, and when I asked Maria, she reassured me that it did not matter, the other community was the same *más o menos*.

While I doubted what was true, there was clearly a general sense of fear in the community. I was concerned that the workshops, including my own, and the films about contamination that the people in the Sápara village watched over and over again were contributing to this fear and paranoia. Maybe it was not as bad as they thought, I thought. A comforting white noise that I had become accustomed to had rendered all other thoughts inaudible: I am rational, I am objective, and I understand. There is nothing to be afraid of. It feels safe here. I feel safe here.

*

I returned from my second trip to the Sápara village on Wednesday July 23, 2014. Though I felt it had been a better experience overall, I still did not have anywhere near enough information to write the report that the NGO required from Maria and me. If I had known it would be my last visit to Sápara territory I might not have been in such a hurry to leave, but I was tired and had begun to feel sick. I was desperate to get back to the comfort and familiarity of the city. I wanted to be on my own, to sleep and to recover, and I needed to gather my thoughts after another "failed" trip.

But as the days passed in Quito following my return, I only felt more tired, and more sick. I developed an unbearable headache, and aching pains throughout my body. My temperature began to fluctuate wildly, and so I slept on bags of ice during the heat of the day, and I took hot showers when I could not stay warm at night. It turned out I had contracted dengue at some point during my trip, and as the fever eventually began to pass a few days later, being in the rainforest felt like a distant bad memory. But while I was sick, I had remembered a girl who was three or four years old crying in the village. She never seemed to stop, and nobody seemed to be doing anything about it. When I asked what was wrong, her mother had told me that the girl had a pain in her head. When I asked why, she told me she did not know. And when I asked for how long she said "A year. Or more". I thought back to the emotional request that the President of the community had made, for help with a medical fund, and I felt glad I had left when I did.

Real-ising Fear

After a week or so, I tried to contact Maria to begin writing our report. I had come to understand that I would be doing all the writing as Maria was not sufficiently literate in Spanish or English to do so, but I still needed her input and knowledge to complete it. Despite the looming deadline, Maria seemed to have disappeared. I called her phone, contacted her via Skype and Facebook, waited for a response that never came, and then tried again. At first, I thought she was avoiding me as she had shown little interest in following up on the report after the workshops, but I later heard that she had gone into hiding.

In 2013, following a protest against oil exploration she had attended in Quito, Maria had been notified of charges that had been filed against her for a long list of crimes. She was not alone in this. Tensions between the Government of Ecuador and those who opposed its policies had resulted in over 200 people being accused of various crimes, including sabotage, obstruction of public roads, and even terrorism. By 2015 these accusations had become so commonplace that when classified documents (which included a long list of human rights defenders, environmentalist activists, Indigenous people, academics, and political opponents of President Correa who had been under surveillance) were leaked from the National Intelligence Secretariat in July that year I sat with three friends and read through it to see if we had been named. Thankfully, I was not on the list but two of the others were. Even if I had been named, I would have had nothing to fear, really, and my friends who had been under surveillance were not afraid either, or surprised. One was a white man from the United States who worked for an NGO, and the other was a middle-class Ecuadorian man who himself now worked for the Government.

But when Maria had mentioned being labelled a terrorist, she had been more concerned. She said while we were eating lunch one day at her home in Puyo:

> They say *I* am a terrorist; they want to put *me* in prison. But they are the ones who use violence, *they* are the terrorists.

On a few occasions, she had told me in passing about harassment and intimidation. But as with the "men in the forest", when I tried to get more details, she would change the subject. I later learned from an anthropologist friend who also worked with her that police had been to Maria's in the night, harassing her and her family, but even then, the details remained ambiguous.

Throughout June and July 2014, tensions between the Government and Indigenous Peoples had been growing once again, partly due to a controversial water law that had been proposed. The approval of the *Ley de Aguas* (Waters Law) in 2014 provoked protests from Indigenous and environmental activists, which included a march across the country from indigenous territories to the capital city. Opponents of the law argued that by centralising control over water resources, the government would gain unrestricted access to key water sources that would be used to facilitate new large-scale mining concessions and the expansion of the agro-industrial sector (see, e.g., Hopfgartner, Moreano, & Santillana, 2016, p. 14). Hundreds of Indigenous people marched from their communities for up to ten days and were then joined by thousands more Indigenous people and other protestors in Quito. I do not know what Maria feared during this time specifically, but she decided to retreat to the cover and relative safety of the rainforest.

On September 10, 2014, having still not heard from Maria, I received a voice message from a close friend in Quito. She sounded panicked and wanted to check that I was okay. She knew I had plans to return to the Sápara village soon, and in the message, she told me that a small commercial plane had crashed in the rainforest near to where I had been. That was all she knew, and she had feared that I might have been on it. I managed to find more information online and learned that the plane was one of the four that Maria and I had used in the months before, and it had been flying from a forest community to the same airport in Shell-Mera. Worried, I began phoning around, trying to find out if Maria or any of her family had been on board, and though I was not able to confirm that she was safe, I eventually learned that the pilot had been alone in the plane.

In the process of reading about the accident, I also learned that it was the sixth time a small plane had crashed in Ecuador in the past two years and though I did not know it yet, three more planes (at least one of which had also been among the four I had used) would crash within the next few months. I took notes on each as I heard about them:

- September 9, 2014: HC-CAB, Cessna 206, *Aerokashurco*. Crashed. Pilot died, no passengers. Sixth time a light aircraft has crashed in the region in the past two years.
- October 1, 2014: HC-COL, Cessna 206, *Aerokashurco*. Crashed. Five died, including pilot. Two survivors severely injured. On way to a ceremony re: human rights abuses in Sarayaku.
- December 28, 2014: HC-CDL, Cessna 182, *Aerokashurco*. Crashed. Four injured or dead? (two nuns, a priest, and the pilot).
- February 12, 2015: HC-CPS, Cessna 206, *Sankip*. Two pilots and passenger survived and rescued.

On a personal level, the time between my last flight into Sápara territory and when I heard of the first plane crash was characterised by beginning to real-ise my own position in my research. I mean this both in the sense of becoming more aware of my role (and the role of my opinions, preconceptions, and biases) in shaping my research and also in the sense of becoming aware that I was, and still am, *really in it*. It was the first stage of the imaginary, artificial barrier between me and my work being broken down, which was instigated by beginning to see abstract risks – of plane crashes or of sickness – as real. The "real" evidence, such as learning about actual crashes of planes I had recently used, or getting sick myself, forced me to look back on the things I had been told and to think carefully about how dismissive I had been. I began to be concerned about the extent to which I had been adding to or perpetuating a process of Othering the people I had been working with, and I began to question how useful my research was, as well as how much personal risk I was willing to take to pursue it.

Feeling Fear

After hearing of the first plane crash, I decided to stop using the light aircraft to fly from Shell-Mera to the "real" Amazon, and I considered alternative ways to return to the community. The only option available, though, was to walk through the rainforest with Maria, which she had previously suggested. She told me she had done it many times and that it was "much better than flying". But trekking for multiple days, or possibly even weeks, through a landscape only accessible on foot and with no way of contacting anyone in the "real" world in an emergency, seemed to me just as dangerous as the dodgy planes. Though I was tempted by the adventure of it, I was not willing to take the risk. It all felt a bit too real now. And so, in the end, I chose to change the site of my research altogether.

This decision, though, did not sit well with me for two reasons. The first was because, despite knowing that nothing had actually changed (i.e., the risks before and after, statistically speaking, were the same), my own proximity to the accidents and my experience of getting sick had shifted my *perception* of the risk. It was not as if I did not know before that small planes could crash or that I could get sick in the rainforest: I had been warned of both, and I had seen Maria's fear of flying and Other people's fear of sickness in the Sápara village. But there was a qualitative change in the way I *felt* about the risks. Coincidentally, or perhaps not, a few weeks earlier I had written a blog post about how being "rational" was a characteristic that I considered to be a part of my identity, and how it might be something that would limit my ability to conduct my research:

> So often we talk about being rational, making decisions based on established facts and existing knowledge, as if it is, and should be, the aim of all people at all times. Ways of being or knowing that sit outside of accepted knowledge can open a person up to being dismissed, discredited, or ridiculed, particularly in the academic world. Anybody who knows me knows that I am a somewhat methodical and "rational" person (most of the time). I love questions and puzzles and finding answers, and I struggle with things being disorganised,

> chaotic, or inefficient. This is probably why I have found beginning my research with the Sápara nation, an Indigenous people here in Ecuador, so difficult.
>
> *(Byrne, 2014)*

The realisation that my rationality was not something I needed to overcome, but had instead been an illusion all along, was not easy to get my head around. The second reason was that when I had observed the same apprehensions in others, I had unconsciously used their "irrationality" to Other them and to unconsciously ascribe to them characteristics of what "being indigenous" meant to me: being irrational, not understanding flying, living in a world that *they* do not understand, and being paranoid, fearing things that are not there. The inconsistencies between my behavioural decisions and the way I imagined myself created a disturbing cognitive dissonance.

Moreover, I found it particularly disquieting that, while I was now taking the risk seriously like Maria, she continued to use the planes anyway, whereas I immediately chose not to. The asymmetry in our relative positions in relation to the work we were doing became more visible to me. Though we both had a "choice", in some sense, regarding the risks we were willing to take, the potential personal costs of those choices were very different. It brought up an uncomfortable question for me: why, and for whom, was I doing my research at all? This question remained with me throughout my research, and it is not one that I have a clear answer to even now. But what became apparent to me was that I perceived my fieldwork and my fieldsite in a way that was quite problematic. It felt (at least to begin with) like an adventure; a voyage of discovery. I was travelling to a different place and to a distinct culture, but I also felt as though I was going to a different time, and none of it was real.

Fabian (1983) identified the role of time in anthropology, whereby the anthropologist and their work are situated (by the anthropologist, at least) in the "here and now", while the objects, the people, and the places contained in the work they produce are situated in the "then and there" (Fabian, 1983, pp. 102–104). The connotations of various terms that are associated with Indigenous Peoples (savage, primitive, pre-capitalist, premodern, undeveloped, under-developed, first peoples, tribes, aborigines, and so on) all also suggest a temporal position in the past. And though "indigenous" is itself an accepted term in the present political moment, it carries with it much of the same meaning: "they" are not like "us" yet, "they" are less than "us". They are Other. Though I felt uncomfortable admitting it to myself, it was the very Otherness of the culture and of the people that had attracted me in the first place. And when I was there, it felt like a window into an-Other time. Each time I left, the memory of it felt fictional because it was not, and is still not, *my* real life.

When my fear became even slightly closer to that of Maria's, which I had dismissed as naïve or an over-reaction, I had the option of avoiding the risk altogether. So, rather than returning to the Sápara village, I decided I would instead accompany Maria and other Indigenous activists on marches, and to conferences, protests, and meetings, and I would spend more time in the office in Puyo. This, I convinced myself, was probably better anyway; these spaces were, after all, the frontline where the "real" fight was happening, while the community I had visited was only distantly connected to the everyday negotiations around indigenous environmental politics. But if I am honest with myself, I know that it was also a value judgement in which I decided that *my personal safety was more important than my research* in the Sápara village.

Regardless of how I rationalised it, in making the choice to not return, I became aware of just how fortunate I was and quickly became ashamed of how readily I had "given up" on something I had previously considered to be so important. I shifted from having the "desire to enter into the world around me and having no idea how to do it" and my main concern being "the fear of observing too coldly or too distractedly or too raggedly" to feeling the "rage of cowardice" and "a sense of the utter uselessness of writing anything" at all. As observed by Behar in *The Vulnerable Observer: Anthropology that Breaks your Heart*, all of these are part of the "voyage" of anthropology, but none are the stuff PhDs are made of (Behar, 1996, p. 3).

So, while Maria was in hiding in the forest, I made my retreat to the relative safety and familiarity of the city. I never got the chance to discuss my new plan with Maria because, while still waiting to hear from her, fear went from being something I had observed in others (and had begun to take somewhat seriously) to being entangled with and embedded within my fieldwork and my experience of it.

Knowing Fear

If they kill us, I hope they
kill her first
So that she doesn't have to
watch me die

Fear is not just an emotion; it can also be a tool. It was described by de Becker (1997) as a "gift" that allows us to keep ourselves safe from danger; an essential part of the intuitive process that helps us to make better decisions. Sometimes what seems irrational, or like a silly over-reaction, is fear telling us to do whatever is necessary to escape danger. But when escape is not possible, when there is no *hope*, fear becomes far less useful.

If they only want our money,
why do they hit me every time I try to speak?
Every. Fucking. Time!

All I want to do is tell my wife it is going to be OK, but I can't, and it isn't. I can't breathe. I can't speak. They won't let me speak in English and every time I try to, they hit me. In the face or the chest, or on top of my head if I curl up. Sometimes just once, and sometimes over and over and over. I lose track when I pass out, and when I come round I am so disorientated – by the noise (it seems odd that the soundtrack to all of this is reggaeton), the smell of body odour, bad breath, and cigarettes, and the taste in my mouth, something bitter burning my lips and my tongue… and blood, I think, and salt, I think, and by realising that I am crying – that I forget what is happening.

I don't know which one it is that keeps doing it, but I think it is the one in the front passenger seat. The big one. I can't see him, now. I can't see anything. They sprayed something in my eyes as soon as they got in. I never cry… My wife is here, she isn't crying, but I can hear her breathing heavily. That's good, I think. She's breathing. I ask her if she is okay. Of course she isn't. He hits me again. Fuck Off! He hits me again. She lies: "I'm okay, I'm okay", and he tells me to tell her to shut up.

How am I supposed to do that without speaking in English?

¡Huevón!

He laughs and he doesn't hit me again… for a while.

We had left a friend's pop-up restaurant in a quiet part of the city a while earlier, just before 11 pm. It was raining heavily, as it often does in Quito, so instead of walking the mile or so home we took a taxi. You do not put your hand up in the air in Ecuador, you put it out to the side like the locals, or they will know you are a gringo and will charge you double. As soon as a taxi pulled over, we had jumped in to find a car that was dirty and smelled of tobacco, with a young and quiet driver who was listening to loud music. At that point nothing seemed out of the ordinary and it was only when the taxi stopped unexpectedly in a dark side street that I suddenly realised something was about to happen. But by then it was too late.

Shit.

I have forgotten more than I remember, now. I have no idea what my wife and I were arguing about on the way to the restaurant, and I do not remember the meal, or for how long our assailants kept us in the taxi. Maybe 2 hours, maybe 3, maybe more? Not that it matters. My wife remembers more than I do. I do remember, though, that once I knew we could not escape, I was not afraid. I was at first. When the three men opened the doors and jumped into the car, sprayed a liquid in my face and pushed their fingers into my eyes, I struggled. When they pulled my hood over my head and started choking me, and punching me, I shouted and fought back. But not for long. It only made things worse. Fear would have been futile. It is a useless tool without hope. There must have been pain, but I do not remember it. I was angry, and sad, and felt guilty for being the reason that my wife was there at all, but I did not feel much else. At some point, I got bored, I think. I thanked them for not hurting her, and I got irritated with them for playing the same stupid song over, and over, and over. And I told them. And they laughed. The rest is hazy. Each time I came to, I had to start over:

We are in a taxi
They only want 'El Money'
They don't want to hurt us
My wife is okay
There is nothing I can do
Is there anything I can do?
Don't speak

When the four men did eventually ditch us on the side of the road, in a gloomy Quito neighbourhood in the early hours of the morning, my head was pounding, my eyes were burning, and my feet were freezing. I realised they had stolen my shoes. We had no idea where we were, and soon we were walking, relieved to be (relatively) safe. It is a strange thought, to feel thankful to be lost with no money, phoneless and shoeless, bleeding, and almost blind in a Latin America city in the middle of the night. We laughed at how absurd it was that the attackers had given us a few dollars to get home. We did not know if it was a sick joke, or if one of them had felt some compassion towards us. But we had no intention of getting in another taxi. I do not remember much of the walk, but I know we eventually found some police who drove us home. It was November 12, 2014, and by the time we got home we had decided I would quit my PhD the next day.

Evidently, I did not quit. Instead, I spent the whole day in a police station in Quito. And the day after that I received the message from Maria that finally ended our work together.[5] Then I spoke to my wife and my supervisors and tried to come up with a plan. It was at this point that I made the decision to travel to Lima two weeks later for COP20, where I would try to start over again. But the experiences above had a profound impact on the remainder of my research, and on the process of writing it.

*

When travelling to and spending time in the Sápara village, I thought I had always taken what people told me "seriously". I knew by now that I was not a neutral observer, but I still felt as though I was outside of the context looking in, and I was able to keep myself at a somewhat safe emotional distance from what I saw and heard. This had been the same throughout my time in Ecuador from when I first visited the country during my undergraduate degree from 2009 to 2010. During that time, I had organised "Toxic Tours" to the edge of the rainforest for students, journalists, activists, and anyone else who had taken an interest in the effects of oil extraction in the Amazon. We would travel to the sites of contamination and visit the homes of *Los Afectados*,[6] where people told us their stories and experiences of sickness, violence, and death. Although the trips invariably began with excitement and intrigue, and a feeling of adventure, albeit with a soberness fitting the theme, many of the people I travelled with would soon be shaken by what they saw and heard. Sometimes they became quiet and withdrawn or were moved to tears, but I was not.

Some seemed to take it in their stride, quickly connecting emotionally with people they had never met in a way that I found impressive and a bit strange. On one occasion I acted as a guide for a photographer who was documenting the lives of the people who lived along the oil pipeline, and so I took him to their homes where we listened to their stories while he took pictures. I was surprised with how comfortable he appeared to be. I always felt awkward, like I was intruding and should not be there, and I remember wondering if it was because he was taking it *less* seriously or *more* seriously than I was. I felt sadness and indignation for the people we met, and I believed what they said, but at times I was sceptical of how "scientifically accurate" their claims were. I took the photographer, Johan Bävman, to the houses of the people affected by contamination along the pipeline that runs 503 kilometres from the Amazon rainforest to the refinery in Esmeraldas on the coast. At one family's home, he photographed a father and son. The caption that accompanied the portrait when it was published read: "Juanito is 17 years old. He is malformed after being born with meningitis. Since the oil companies came to the village, several children were born with deformities" (Bävman, 2010). I remember wondering at the time whether oil contamination really could cause meningitis, or whether people had just become so afraid of contamination that they thought it was the cause of all illnesses.

In those fearless days I also visited the home of a man I had been working alongside in Quito, Pablo Fajardo, the lawyer for the plaintiffs in the case against Chevron. While we sat drinking coffee in his garden, he told me about his brother being murdered in 2004. He had been tortured before he died. Nobody knew who did it, or why, Pablo said. But there were rumours that it was a case of mistaken identity, and that Pablo was the real target (Páz y Miño, 2008). I always listened to every detail, and I cared. I was not desensitised nor was I sensitised. I was reasonable and rational, and I was curious. I was never afraid, not *really*, because it was not real. I didn't feel afraid even when I had a close call in that first year. I was out running one afternoon when three men attempted to pull me into a van in a public square in Quito. Nobody knew who did that either, or why, but suspicious-looking vehicles and people had been hanging around outside our office, and I had been warned to be careful. I filed a *denuncia* with the police (another to add to the list of reports that the staff from the office had made), and I filed the experience away as another story to tell. It was just intimidation.[7]

Fear in the Field

In the days, weeks, and months following the incident in the taxi, two changes in my research gradually came about; my perspective on the experience of Others changed quite dramatically, and I began making decisions regarding where (and with whom) I worked differently. As noted above, the shift began earlier with the decision to not return to the Sápara village, but it is difficult to write about this in a linear fashion because it did not happen that way. Instead, there were two things happening at once: one was a retrospective reimagining of the encounters I had had so far with Indigenous people, and the other was focussed on finding a path I could follow in order to continue my research. These simultaneous processes felt as if one was situated in the past and the other in the future. But in the present, it made for a chaotic and at times incoherent thought process, which was contaminated by experiencing intense and often contradictory emotions, including a feeling of *real* fear that I had never known before. I use the term "contaminated" here in reference to Tsing's reimagining of "contamination" as "collaboration". What felt like my rationality being contaminated by emotion, first by the fear of Others and later by encountering violence myself, could just as well be understood in terms of collaboration and I eventually came to think of it that way:

> We are contaminated by our encounters; they change who we are as we make way for others. As contamination changes world-making

> projects, mutual worlds – and new directions – may emerge. Everyone carries a history of contamination; purity is not an option.
>
> *(Tsing, 2015, p. 27)*

But at the time, I did not think about it much at all. I just *felt* it. I was afraid of things that were there, and of other things that were not. Sometimes it felt intuitive and hyper-rational, allowing me to see risks I had been oblivious to (or had ignored) before, while, at other times, it felt paranoid and irrational, causing me to experience intense physical and psychological responses to danger, even when it did not exist.

Ultimately, though, this disordered and distorted time led me to a different understanding of my research, my own place in it, and its epistemological limitations. It is, as Jamison (1996, p. 203) put it in *An Unquiet Mind: A Memoir of Moods and Madness*, "an awful prospect, giving up one's cloak of academic objectivity". And it is even more awful when it is not a choice: it was as if the cloak had been pulled away, allowing subjective experience and emotion to pour into and out of my work, forcing me to concede that "of course, my work *has* been tremendously colored by my emotions and my experiences" (Jamison, 1995, p. 203). I had acknowledged this in theory long before, including in my research proposal, noting how important it is for a researcher to reflect on one's own position in relation to their work and to the people they interact with, and to consider how this might shape the research that is produced. Yet I had maintained a paradoxical notion in my thought process; that acknowledging my irrationality and subjective position was a way to *overcome* them and become *more* rational and *more* objective. I was treating them as challenges or imperfections, rather than as additional components (or collaborative contaminations) of a perspective that is always subjective and is always partial.

Though I felt alone in this time, I was not. Years later, as I was finishing writing my PhD thesis, my wife, Jo, was required to give a presentation on the theme of trauma for her work in Children's Social Care. She asked if she could practice the presentation with me, and in doing so she was able to say things to me that she never had before. The following passage is an excerpt from her presentation:

> It seemed wrong to allow these criminals to ruin George's research and potentially his whole career, so we stayed. I know now that was the right thing to do, but we lived in terror for the next eight months.
>
> I got through it the only way I knew how; I made my world really, really small. I barely left the house, and I isolated myself from everyone around me. I suffered insomnia and became addicted to the sleeping

pills that were sold unregulated over the counter. And I ate as if my life depended on it.

I didn't realise at the time, but because my brain had already been affected by my childhood experiences, it made me more susceptible to the trauma from the kidnapping. Somebody else could have experienced the exact same thing as me and have come out mentally unscathed, but I was already vulnerable.

George and I were both so proud and unwilling to show vulnerability that we kept what had happened to us a secret and struggled through life, almost allowing what happened to break us up.

That night after she practised that presentation, and in the weeks and months that followed, we began to really discuss our experiences in a way that we had not before. We did not talk about it at all in the beginning, or for a long time after, other than quite superficially. Though the experience was shared, the time that followed it was characterised by silence and secrecy. As I watched Jo become more and more withdrawn, I wanted to make my world smaller, too, but I couldn't. I had to make it bigger and find somewhere else to do my research.

*

I arrived in Peru, less than three weeks after receiving the message from Maria, which had put an end to our work together. This meant I had to find an alternative way to conduct my research and to collect sufficient data to write a thesis. Doing so was no longer motivated by a curiosity about the subject, or a want to do something "good", or even by a self-centred desire to be awarded a Doctorate. My primary motivations were fear and guilt. I was afraid of staying and failing, or of leaving and quitting, and I was wracked with guilt for having put my wife through an experience that had put her physical and emotional health, as well as her life, at risk. I was also aware that, although Maria had ultimately been the one who ended our work together, I had already decided that I would not return to the Sápara village. Knowing that I would not go back and that any work I produced would have no positive impact on the people I had visited made writing about it at all seem pointless and doing so would be purely "academic". I was also aware that my fleeting visits, during which I extracted some data, took some pictures, and drank some *chicha*, had done nothing to challenge the pattern of exploitation of Indigenous Peoples. I just hoped I had, at least, "done no harm". At this stage, I felt as though I had to continue and complete my research so that something "good", or at least *something*, would come out of it all. I wanted to make it better, but I was so tired.

Lima, where COP20 was held, is a sprawling and overwhelming city: hot, loud, and uninviting. Or that is at least how it felt to me, having not yet begun to deal with the emotional impact of recent events. I spent the first few days avoiding the heat and the noise, avoiding work, and avoiding people. I left my hotel only to buy food, though I barely ate, and to buy sleeping pills, though I hardly slept. I saw risk and danger everywhere, and whether it was "real" or "imagined" did not matter. I had become suspicious and dismissive of people. And so, I closed myself in, away from everyone and everything for as long as I could. My memories of Lima are sparse and disjointed, muddied by emotions and by self-medicating to control them. My notes are equally scant, and many of the photos I have feel as though they were taken by someone else. In writing about it I find myself lost in the city again, both in the memory of the feeling, experiencing it over, and in searching for what in this memory is "true" and "real". In *Writing Selves*, Perreault cites Audre Lorde's diaries, where she reflects on the apparent impossibility of writing in the present what was once "real"

> The act of writing seems impossible to me sometimes, the space of time for the words to form or be written is long enough for the situation to totally alter, leaving you liar or at search once again for truth.
>
> *(Lorde, 1978, p. 52 cited in Perreault, 1985, p. 20)*

There are large sections of my fieldwork that seem so distant in space and time, now, that to write anything "factual" about them would be a stretch, to say the least. But I know that I learnt a lot, even if I do not remember exactly how and when. "You will remember what is important". I forgot that I had been told this by the young Sápara woman when I visited the village,[8] until I saw it in my notes while writing up. What I remember from Lima is feeling, for the first time, that the things I had been reading about, talking about, and writing about for several years were real. The people and places, the risks, and the violence were no longer just abstract thoughts and subjects of study that existed in articles and books or in my field notes. I remember now, though I did not then, something Smith wrote about "re-membering" in *Decolonizing Methodologies*:

> … re-membering in terms of connecting bodies with place and experience, and importantly, people's responses to pain. While collectively indigenous communities can talk through the history of traumatic events, there are frequent silences and intervals in the story about what happened after the event.
>
> *(Smith, 1999, p. 147)*

For better or worse, my body felt newly connected to place and experience, and I began to think differently about other people's responses to pain, trauma, and fear. I tried to put back together the events that had led me to Lima, and particularly what had happened with Maria, who I had been told would also be at the conference. I thought back to her "irrational" fears of flights and the "men in the forest", and I considered her retreat to her territory, which had apparently been the result of her feeling under threat. Her tendency to avoid answering questions or to brush over particular subjects had been confusing and frustrating for me, and the disjointed and fractured accounts of fear by the people in her community had left me doubting what had been true. I could understand all of this better, now. In a sense, I came to see silence or ambiguity as a means of survival in a world that is, according to Tsing, characterised by vulnerability and precarity

> Precarity is the condition of being vulnerable to others. Unpredictable encounters transform us; we are not in control, even of ourselves. Unable to rely on a stable structure of community, we are thrown into shifting assemblages, which remake us as well as our others. We can't rely on the status quo; everything is in flux, including our ability to survive.
> *(Tsing, 2015, p. 20)*

But still, being accused of being just like everyone else and of being there only to take from them had left me feeling confused and resentful, particularly as I felt I had put in considerable time and energy for little in return. But Maria's caution or mistrust of me, if that is what it was, had been quite rational. I had begun my research thinking that *me* learning about Indigenous Peoples would benefit *them*, and that through my writing I might be able to share their voice or advocate for them, or something.

*

In *An Open Letter to a Young Researcher*, Adams (1979, p. 451) reflected on the history of a particular Senegalese community's experience of encountering Europeans (a category in which she includes North Americans). She recounted a visit from a young French researcher (yet another passing European), and though she was not sorry for receiving the student coldly, she regretted not explaining to them why. There were many reasons, some were made explicit while others were implied in her story, which described various bouts of "research" that had been conducted and the repeated attempts at "development" that had been made by "experts" in the region. The experts, she noted, were all from outside, and she argued that the very notion of being an expert is implicitly linked to being an outsider:

> Whatever the level, the expert's function is always the same. Even the term "expert" involves hyperbole. None may be an expert in his own country; it's an expatriates' title. In Britain, a doctor is a doctor; he'll be a medical expert if he goes to help halve the birth rate in Bangladesh. (Or, just possibly, if he's asked to join a panel on slum clearance in Glasgow).
>
> *(Adams, 1979, pp. 473–474)*

Reflecting on "research", Adams states that the experience of the community had demonstrated that "Researchers aren't looking for the truth" (Adams, 1979, p. 471), and instead that the disciplines of history or social anthropology have served to deny the present existence of the people and places that are "researched":

> ... the practice of history as a discipline denies Jamaane present life, because it covers only the period between contact and conquest. I am saying that the practice of social anthropology as a discipline denies Jamaane present life, because it can speak of Jamaane only in an intemporal present tense which is a disguised past tense; or, more straightforwardly, in the pre-capitalist these disciplines, whose claim to honour is that they recognize African realities, add up at best to a partial truth that covers many lies by omission, and this is what makes it possible for Africanists to countenance the practice of development experts and the very existence of the category "development studies", which makes nonsense of everything they might be thought to have aimed for. I write intemperately because I am angry.
>
> *(Adams, 1979, p. 479)*

I had not read this piece (or anything like it, really), which was written over 40 years ago, until after my doctoral research, despite having spent over a decade in development studies and anthropology. This is indicative of a failure of myself and of the disciplines more generally to endeavour to take reflection and self-critique seriously. The crisis identified long ago by Lewis (1973, p. 581) – that in "the field" and in the classroom there is an estrangement between the anthropologist and the "non-white" people they study – still exists. I had *tried* to work around this. Yet somehow, despite all the reading and thinking about "decolonisation" and the "homework" I had been doing, I still thought *I* was *helping*. Even in her angry account, Adams Adams - who lived for many years in Senegal and worked alongside (as well as wrote about) local farmers defending their vision of "people centred development" against incursions by the State development corporation (Adams, 1979) - does not quite seem ready to position herself within that which she is so angry about and instead rejects the identity of "academic" altogether.

Likewise, I recognised that *my* identity carried with it symbolic meanings for others, and that these could potentially be problematic, but I still did not understand why Maria did not trust *me* because I perceived myself to be different. The last line in the first quote from Adams above ("if he's asked to join a panel on slum clearance in Glasgow") perhaps alludes to part of the reason why I find the idea of being lumped in with other "development experts" and "academics" so unpleasant; as well as denoting "expatriate", these titles can also be symbols of class privilege.

As I pointed out in my methodology, I consider myself to occupy a space that leaves me an outsider in relation to my research subjects but also in relation to the academy and the history *of* academic research. This is because my relationship with the academic world, largely because of growing up in an underprivileged household and having parents who were poorly educated, has been quite antagonistic. I see the potential value of academia as a tool of social mobility at the personal level, but at the social and cultural level I also consider it to be an apparatus of subjugation that creates and maintains economic, social, and epistemological hierarchies. I have always been suspicious of and resistant to authority, and particularly to the cultural dominance of one group over another, and I resent having to confront the idea that *I am, in many circumstances, in the privileged position*. And yet, as a white, Western man, and as an academic researcher, I must do so: to "clean up my act" it is necessary to engage all aspects of my identity (Rich, 1984, p. 122).

From the beginning, Maria had seemed suspicious and dismissive of me. She certainly did not trust me, and eventually she was, I think, angry with me. I came to understand that better, and I began to wonder if she might even have been afraid of me, too. Probably not, but I had begun to see and to feel fear everywhere, and I was also angry. Any semblance (or pretence) of objectivity was gone, and it was particularly difficult to admit to myself that helping was not really the reason I was doing my research at all. But selfish fieldwork is nothing unusual, even if many researchers would prefer not to admit it. As Barley put it:

> Frankly, it seemed then, and seems now, that the justification for fieldwork, as for all academic endeavour, lies not in one's contribution to the collectivity but rather in some selfish development.
>
> *(Barley, 1983, p. 9)*

I have become quite sceptical of the role of fieldwork in the production of knowledge, and particularly the impact of fieldwork on the people (both those researching and those being researched) who are involved in it. And yet I am also aware that there is a methodological contradiction lurking here: fieldwork might be difficult to justify in terms of contributing to the collectivity, and it is certainly hard to claim that it necessarily benefits anyone other than the researcher, *but it was the very experience of fieldwork, including "failed" fieldwork, that allowed me to reach a better understanding of how problematic it is.* In particular, losing objectivity and becoming more intuitively aware of my position and motivations was precisely the reason why I began to appreciate, in a more connected, bodily sense, the fears of Others.

Re-Thinking Fear

"Objectivity", or at least the appearance of it and the *feeling* of it, is a partial perspective that positions and values one view (i.e., that of the "rational" observer) above all others. As Haraway (1988) pointed out, it is the preserve of those who find themselves in a privileged position of power. It is only the unmarked category that can claim this power "to see and not be seen, to represent while escaping representation". The underlying assumptions, concealed within such assertions of being part of some unmarked objective category, are often implicitly being Man and being White (Haraway, 1988, p. 581). Objectivity is treated as synonymous with "reason" or "rationality", and as a prerequisite for seeking "truth", while a "lack" of it is viewed as an epistemological shortcoming. Conversely, then, the perspective of the Other can never be treated as objective or rational, unless one accepts the limitations of one's own partial and subjective perspective. As Janack (2002, p. 268) noted, however, the connection between truth and the epistemic virtue of being (or trying to be) objective does not mean that "failures" of objectivity are *always* or *only* epistemic failures.

Hegeman and Stocker (1996, p. 119) considered the role of emotions as complex and highly nuanced, and they note that subjective experiences of fear and anxiety can engender both epistemic failures *and* successes; they can both aid and inhibit. If being subjective, irrational, and emotional are characteristics that tend to be associated with less dominant groups (Indigenous Peoples or ethnic groups Other-than-white, women, or "the poor" and poorly educated), and if these characteristics represent an alternate and partial (but not "worse") perspective, then it follows that the supposedly objective, rational, white men are at least missing something. A particularly interesting observation relating to this is that there is evidence that people from marginalised groups are "generally better at reading moods and interpreting facial expressions than are those in power, even when those in power are the very people being interpreted" (Hegeman & Stocker, 1996, p. 119). Also, studies of risk perception have suggested that some groups or categories of people experience emotions and see situations differently to others. For example, Finucane, Slovic, Mertz, Flynn, and Satterfield (2000) noted how men in general perceive less risk than women, but that "non-white" females and "non-white" males differ very little from one another in their perceptions. The apparent gender difference appeared to be driven by a lack of perceived risk among white men in particular, with around 30 per cent of the group experiencing extremely low-risk perception (Finucane et al., 2000, p. 159).[9] This suggests that it is not race or gender per se that underlies the disparity, but rather that white men perceiving less risk is caused by something else.

More recent studies have found that this disparity holds even when age, income, and level of education are accounted for, and this is not only for risks that one might imagine to be specific to certain groups (such as sexual violence against women, or state violence against ethnic minorities). Macias (2016) found that in the United States perception of "risks once thought to be more equally distributed throughout society such as climate change and nuclear power plants" were found to be significantly greater "among people of color than among members of the majority [i.e., white] population" (Macias, 2016, p. 126). The study offers as a possible explanation, which is that inequalities in environmental impacts might mean risk is more palpable for those who are more marginalised because they bear a disproportionate share of the negative outcomes, which they do. But this does not account for the disparity holding true across socioeconomic status, political views, and rural/urban residence. There is a danger of making essentialist assertions about gender and race differences here, and Macias comes uncomfortably close to this by asserting that part of the reason for this disparity is that "environmental values among people of color persist over generations", without qualifying from where these values originate and how they differ from those people who are not "of colour". But a relatively simple and non-essentialist explanation for this might be that those who perceive more risk do so because they have a shared experienced a more traumatic cultural past and present. Historic trauma can be understood as both a cluster of traumatic events and as a disease itself to which there is no single cultural symptom or response (Smolewski & Wesley-Esquimaux, 2004, p. iv). Alexander (2004, p. 1) discussed the complexity of the term and the potential impact of "collective cultural trauma", which occurs:

> ... when members of a collectivity feel they have been subjected to a horrendous event that leaves indelible marks upon their group

consciousness, marking their memories forever and changing their future identity in fundamental and irrevocable ways.

(Alexander, 2004, p. 1)

Of course, people who are part of communities and groups who have *not* had immediate experience of trauma can have knowledge of risks, and they (we) can have some degree of empathy for those most at risk. But the emotional connection to potential future trauma, the *feeling* of fear, is different in some sense. Take for example this analogy, which highlights the qualitative difference between *knowledge of risk* and the *feeling of fear*, brought on by experience:

> Before I slipped on the ice, I believed it was dangerous to walk on the ice and I wanted to avoid these dangers. But I had only an intellectual appreciation of those dangers and a pro forma desire to avoid them. Having slipped, and without any change of my beliefs or desires or values, I am afraid. It cannot be held that I have new beliefs, such as "I *really* could slip; it could *now* happen to *me*." For I already believed that. To be sure, I did not take it *seriously*, in the sense of taking it with fear.
>
> *(Hegeman & Stocker, 1996, p. 47)*

To take risks seriously in the sense of "taking them with *fear*" is a useful choice of phrase, here, as it draws a distinction without casting a judgement on which assessment of the risk is better or worse. When I consider this in relation to my own experience, I can see how "taking risk with fear" neither clarified nor obscured the "truth", it only shifted my perspective on it. For example, though I had always empathised with the pain, suffering, and fear of Others, experiencing violence and fear myself brought about a qualitative change in the way I felt that empathy. If I had read this written by someone else only a few years ago, I would have seen it as melodramatic, but death shifted from being something I could *imagine* to being something I had, at a specific moment in time, perceived to be real and imminent, and this allowed me to take risk (both to me and to Others) in general more seriously, as in, *taking it with a sense of fear*.

This became apparent to me during the first event I attended in Lima where I learned that a Shuar man named José Tendetza, an Indigenous activist who was expected to be at the conference, had been murdered in Ecuador. I had heard his name in passing before, but all I knew of him was that he had been involved in resistance against the *El Mirador* copper mine, which had been opened in his territory. The many different forms of violence against Indigenous Peoples and the pervasiveness of violence in the name of development,[10] as well as in the name of sustainability (see e.g., Dunlap, 2017), meant that there was nothing particularly unique or shocking about this story. It was not even the first murder of an Indigenous activist I had heard about in the past few weeks,[11] but it felt different this time. During COP20, the murder of Tendetza came up in many conversations I had with both Indigenous and non-indigenous people, and the details changed from day to day and from one person to another. At first, I heard he was on his way to the conference when it happened, then later that he was going to a protest or a meeting about the mine. People speculated on what had happened and who was responsible: The Chinese mining company? The Government of Ecuador? Other Indigenous people who wanted the mine to go ahead? Everyone seemed to have suspicions, but nobody knew for sure. He had been tortured, people said, and his mutilated body was left in an open grave, found four days after he had disappeared. He was drowned, he was shot, or both. Some people thought it was intimidation or a warning to other activists.

On December 8, 2014, there was a side event at COP20. It was a "Public Hearing on Indigenous Peoples' Concerns and Perspectives" and was held at the Lima Museum of Arts (MALI). Once again, the murder of Tendetza was discussed. His picture was projected onto a screen, while people from his community spoke of their struggles, their fears, and of the violence they faced. Many other people gave their testimonies, too, which included various accounts of contamination, land grabbing, and violence. I do not remember exactly what was said, and I did not take many notes. I remember what was important, I think, or at least what was important to me. The Indigenous people who recounted their stories there, and at other events I attended, were afraid: of climate change, of contamination, of governments, and of corporations. I had heard it all before, or some version of it, but it *felt* different. It is not that I listened more intently, or understood their experiences better, or even that I felt more empathy, I just felt sick. I used to find it *interesting*. But my objectivity, rationality, and resilience were all gone. I did not want to hear it anymore.

There was also the play (mentioned in the previous chapter) that depicted a typical interaction between Indigenous Peoples and their Others: people who wanted access to *their* knowledge and the resources in *their* territory. It was the only part of the event where I took notes and pictures. A particular piece of dialogue, spoken by the daughter of the "Wise Man" to "Mister Doctor Professor Koch", is relevant here:

> ... we think of time and space in an entirely different manner to city people. And if we have to live in the city, we are done for, our culture will be lost. Is that what you want? You think that you are better than us, and that we must live in the same way as you; the way that will lead us to damnation. No, ma'am! I have been to school, and my parents

> and grandparents have taught me about my culture and traditions. What you are explaining is not good for us in any way. What is happening here is that you are simply sacrificing us, Indigenous Peoples, for what you call "development". But it is this "development" that leads to destruction. Don't you understand that?

I think it was around this time that I thought back to what Maria and her partner had told me about the Government, *los cólonos*, *los petroleros*, and the elusive "men in the forest", and I also reflected on how they might have seen *me* as part of the same process. I did this at some point: maybe then, maybe later. The details and the source of the danger were always vague, which had made me doubt the truth and validity of their story, and my doubtfulness may have contributed to the silence and vagueness. But now I think that perhaps the type of fear they described, the need to stay vigilant of an ever-present but indefinable threat, is the most toxic and the most "real" of all. As Bauman (2006) put it in the opening pages of *Liquid Fear*:

> Fear is at its most fearsome when it is diffuse, scattered, unclear, unattached, unanchored, free floating, with no clear address or cause; when it haunts us with no visible rhyme or reason, when the menace we should be afraid of can be glimpsed everywhere but is nowhere to be seen. "Fear" is the name we give to our *uncertainty*: to our *ignorance* of the threat and of what is to be *done* – what can and can't be – to stop it in its tracks – or to fight it back if stopping it is beyond our power.
>
> *(Bauman, 2006, p. 2)*

My memory of that time is as diffuse and scattered as the fear I was experiencing. What I have come to realise through trying to piece my own memory back together, re-membering, is that the truth of the memory and the reality of the threat as understood by someone else are, in the first instance, unknowable, and beyond this they are irrelevant if the intention is to take the subjective experience and knowledge of Others seriously.

When people heard about what had happened to my wife and I, they often responded by asking questions about the details:

> *Didn't anything seem suspicious?*
> *How many men were there?*
> *How long did they keep you for?*
> *Wasn't there anything you could have done?*

As if these details mattered.

Sometimes they suggested what we *should* have done (or worse, what *they* would have done) differently. And then each time I was required to tell the story – to police, to friends, to colleagues, and to my doctor, and psychologist after I was diagnosed with Post-Traumatic Stress Disorder – the details changed and became increasingly vague. Though people listened, I often felt as though they did not hear – and certainly did not feel – what I was trying to communicate. I became frustrated with the feeling that I had to justify my actions, my experiences, and my emotional responses to them. I began to anticipate questions, answering them with what was expected and what would not be challenged, rather than being honest and telling them my truth: I felt as though even when I was being asked to speak, I was being silenced. I gradually became aware that throughout my research I had unintentionally been asking the same of Others, requiring them to speak of their experiences of violence and trauma in ways and at times that suited me. In a study of *History, Violence and Mental Health in Ecuador*, Capella, Jadhav, and Moncrieff (2020) described "forgetfulness as a coping mechanism" in such situations. I had not previously considered this, partly, I imagine, because I had only ever been the one 'giving' (rather than being given) voice. On the question of why communities are reluctant to talk about their violent histories, Capella, Jadhav, and Moncrieff cite linked arguments made by Du Bois (1903), Fanon (2004), Freire (1970), and Goffman (1990) and hypothesise that

> ... communities might find it uncomfortable to openly address post-colonial feelings of alienation, disvalue, anger, conformity or dependency and to disclose the historical origins of their cultural identities, especially if these are disvalued.
>
> *(Capella et al., 2020)*

Being required to speak of my own experience and not feeling heard helped me to better understand, or at least gain some insight into, how I and my research were positioned within relations of power and (mis)trust, and how this could cause silence.

In 2016, after I had returned to the United Kingdom, I was told that a female member of Maria's community had been found naked and mutilated, sprawled on the forest floor among discarded machetes and a strange collection of forest cuttings. Some of the Sápara women who were opposed to oil extraction had read the cuttings as a kind of "hit list", and they believed that they were the next targets, while some community members claimed it was suicide, domestic abuse, or shamanism. The perceived link between the brutality against the woman who was murdered and the fear of outsiders, and particularly the oil industry, as with the speculation about the mining company following the murder of Tendetza, illustrated the psychological and cultural impact of the ongoing "war" they say they are confronting (Ofrias, 2018). NGOs that work with Maria have claimed that, like the murder of Pablo Fajardo's brother, it was a case of mistaken identity, and she was in fact the target of the attack. Later, videos began to circulate on Facebook that showed men from other indigenous communities threatening to murder Nema Ushigua, the first female President of the Sápara people, describing in detail how they planned to do so (Tegantai, 2018). Like the plane crashes, learning of this reminded me of how naïve I had been, and I regretted how, while in Sápara territory, I had been inclined to minimise and rationalise other people's fears. It also reminded me why I was glad I had not returned.

In Ecuador's Oriente, fear is a part of life that people seem, at least to me, to have become quite accustomed to. Being afraid of flying or contamination, of the "men in the forest", or of sickness and dying is not irrational at all. It demonstrates a clarity of thought and an understanding of the risks and the very real dangers that exist, as well as the extent to which they can (or cannot) be avoided. Similarly, Cepek (2018, p. 92) reflected on his experience of changing perceptions of risk while working with the Cofán people in North Eastern Ecuador, stating that experiencing "the fear of *cocama*[12] capture helped me to understand how deeply Cofán people associate newcomers with violence and death" and cited *Grief and a Headhunter's Rage*, in which Rosaldo recounts how the traumatic experience of his wife's death during his fieldwork allowed him to understand the relationship between rage and grief. Cepek explained that

> … nothing prepared me to appreciate Cofán stances toward *cocama* until I felt the fear the outsiders brought to my own life. Cofán people were the ones who drew me into their anxiety and apprehension. It was them, not me, who decided that Manuel wanted to capture me. Given their history, their fear made sense.
>
> *(Cepek, 2018, p. 92)*

Cepek considered himself to have been drawn into the anxiety of others about kidnapping, and I experienced a similar contagion of fear. And though he acknowledges that their fear "made sense", he still felt the need to contact Randy Borman (who, though he is Cofán, also happens to be a white man) to check if he and his colleagues "were being irrational". In the past, I have done the same thing, minimising my own "irrational" fears and those of Others, and seeking the reassurance of other "rational" (in my case, non-indigenous) people, despite all evidence suggesting that the risks, including kidnapping and other forms of violence, are real.

The role of fear, and particularly the fear of death, in shaping the way that people interact with the world and with each other was theorised by Becker (1975) whose book *The Denial of Death* argued that

> ... the idea of death, the fear of it, haunts the human animal like nothing else; it is a mainspring of human activity – activity designed largely to avoid the fatality of death, to overcome it by denying in some way that it is the final destiny
>
> *(Becker, 1975, p. ix)*

This notion – that knowledge of death, the fear of it, and the wish to overcome it are the 'mainspring' of human activity – is a compelling thought, and more recently experimental research into the ways in which being reminded of one's own mortality affects one's actions has found strong evidence to support Becker's claim. Greenberg, Pyszczynski, and Solomon, who developed "terror management theory", found that when individuals are asked to consider their own death and are then asked to make decisions, the effects are significant and measurable. The first study they conducted was with municipal court judges in the United States, half of whom were asked to consider themselves dying and the other half not. When asked to set bonds for a crime following the exercise, those not reminded of their mortality set a value of $50, but those, who had considered their own death, set it over nine times higher at $455 (Solomon, 2015). This came as a surprise to the researchers, and it led to numerous further experiments that have tested various ways of considering death and the impact of doing so on the subsequent actions of participants. The experiments have consistently yielded similar results (see, e.g., Greenberg, Pyszczynski, Solomon, 2015; Darrell & Pyszczynski, 2016). What is particularly surprising is that the judges, like most other participants in the experiments, tended to deny that their decisions had been affected by mortality salience: they considered themselves to be acting "rationally".

Considering climate change in these terms is helpful when trying to see the risk of it more clearly (as opposed to more rationally). Though "we" in our privileged "objective" position might well *believe* that there are real dangers and want to avoid them, our appreciation of the risk is, in general, intellectual. It is a partial perspective that lacks an emotional connection, treating the risk as an abstract notion. Meanwhile, those who know and have experienced a history of cultural decimation, violence, death, and contamination need not be in any sense fundamentally "different" in order to take risk more *seriously*, in the sense of taking it with fear. On a personal level, I know this to be to be the case because it was the experience of fear, and the salience of my own mortality, that enabled me to begin to appreciate, in an entirely different way, the fear of Others, and it required no change of my knowledge or my beliefs, desires, and values.

For many Indigenous Peoples, including in the Amazon region of Ecuador, climate change and mitigation strategies, oil exploration, and incursions on their land and territories by outsiders are all often entangled in an overarching fear for their physical survival as individuals and their cultural survival as Peoples (Reed, 2011, p. 526). Research conducted in a Shuar village, Tsuntsuim, in Ecuador, provides a salient example of how the effects of trauma permeate and linger in communities. The report found that when half of the population was displaced in order to make way for a mining project, nine months later almost 40 per cent of those affected showed signs of ongoing mental health problems relating to the trauma (Solíz, 2017). This isn't a unique case: Indigenous Peoples' lived experiences of trauma and the "utter normalisation" of violence against their communities (as illustrated in many works including Green's ethnographies in Guatemala, 1994; Alaska, 2012) permeate the "social memory", leading to a climate of fear, distrust, and silence. Taussig (2003) described the omnipresence of fear in the fieldsites of his research in Putumayo, Colombia, a place that he described as a "space of death", created by colonial extractive industries (particularly rubber) that also engendered the death of communal memory and objectivity (Taussig, 1984). By looking back on his early field diaries, Taussig also reflects on how quickly the memory and connection to fear fade when the researcher leaves the field:

> History lies in the adaptation of materials to time, to the exigencies of life, much as a door handle loses its shine or the keys on a keyboard lose their lettering. What catches my eye reading these old notes for 1970-1972 is the fear of thieves and of violence. This fear is everywhere like the beating of your heart night and day. If it stops, it means you've stopped.
> *(Taussig, 2003, pp. 135–136)*

Since returning from my fieldwork, my own fears have certainly faded: the Otherness of the place in which I experienced it most acutely has allowed it to reside primarily in the "then and there", rather than the "here and now". But I do carry fear, as Tsing put it (2015, p. 27), through a history of contamination: objectivity, like purity, is not an option.

Fear and the Other

What I learned from all of this is that without fear, risks can become distant abstract ideas with little salience: they do not seem real. Thus, if you find yourself about to get into a taxi with a friend who happens to have once been attacked and robbed in one, and they do not want to get in because "something doesn't seem right", you would do well to listen to them and wait for the next one. Even if they are wrong, *the cost of not listening is potentially great, while the cost of deferring to their experience and increased perception of risk is far smaller*. On the same note, if you are ever in the forest with someone who knows it far better than you, and they tell you that there are dangerous men amongst the trees, you ought to *believe them, even if you feel safe*. And if the most marginalised people in society are asserting that not enough is being done (and what is being done is not working) to address an imminent and potentially catastrophic global issue, it would be advisable to take their fears seriously. Those, who perceive risk differently to the white man's supposedly objective perspective, are far from "irrational": light aircraft do crash regularly, Indigenous people do get attacked and killed by strange "men in the forest", and environmental damage, contamination, and climate change are not development "opportunities" presenting "co-benefits" upon which to capitalise. The problem is that it is difficult to appreciate the fears of Others, or the gravity of the risks, unless fear has been felt, experienced, or embodied for oneself.

Something I related to in Cepek's account of fear is that – after retreating to the relative safety of the city in Quito (following Borman's advice that "you can never be too careful") – he returned to his fieldsite because, although he was afraid, he still wanted to finish his research (Cepek, 2018, p. 90). I wanted to finish my research too, but I was not quite so able to rationalise and supress my own fear. And besides this, the cities now felt just as dangerous as the rainforest, and the people I had been working with no longer wanted to work with me. Once my acute experience of danger began to subside, I was able to convince myself, "rationally", "objectively", and "sensibly", to stay and complete my research, albeit in a conveniently reconceptualised way. Thus, I found myself six months into my fieldwork with little to show for it and feeling that I had extracted almost nothing of use to me. I had written a report for the IPCCA that would never be published, and even that was based on methodologically problematic workshops. The shift in my research focus began here. I needed, I realised, to work with Indigenous people who would cooperate with *me*, and I wanted to capitalise as best I could on the situation I found myself in, while also trying to minimise personal risk. And so, I began to search for "them": an Optimal Other who would lend my work a degree of legitimacy without my having to go too far "out of my mind". I did so in a place where *I* would feel more comfortable, and which was not indigenous at all.

The very first event I attended once I managed to leave the hotel room in Lima illustrates this: it was an activist gathering in the Casa Activa, which had been created by a Bolivian organisation called Tierra Activa.

It was a space for various activities aimed at solidifying and connecting the activist movement around climate change, and it claimed to provide "an inspiring vision of what a sustainable future could look like through self-organization" (350.org, 2014). It was not inspiring to me, but it did feel safe. I had imagined it would be a good place to meet and spend time with some indigenous groups or activists, take some notes, take some photos, quote them in my research, and so on. But I was immediately struck by the prevalence of seemingly middle-class, mostly English speaking people who I read as white or mestizo (or at least non-indigenous). It reminded me of socialist/activist youth spaces where I had spent time years earlier in Britain and Australia. In turn, this reminded me that I was gravitating towards spaces that were more familiar and comfortable and where I did not stand out, and also where I was unlikely to learn anything of academic value about Indigenous Peoples' experiences. Like me, the people in attendance were in Lima because of the conference, they had the resources and the time to fly there, and represented, to me, the face of "legitimate" resistance at COP. They appeared to be enjoying themselves. Resistance is fun when it is safe.

Despite the whiteness and safeness (for me) of the space, it was here that I first learned of the murder of Tendetza and began to consider what it might mean for Indigenous people to *be* at COP. I thought about those who had travelled to Lima and what it might mean for them to be there – away from the safety and security of their territories, homes, families, and communities – and what other options they have. I felt like *I* had no choice but to be there, and I resented it. But for me, the cost of being at the conference was primarily emotional, and it was at least a familiar environment in which i could cope with the difficulties I had recently encountered. For the Indigenous people in attendance, the costs were potentially far greater. In addition to travelling to and operating within a space where Western ways of being are given primacy, the historical and contemporary patterns of domination (i.e., colonialism, contamination, and structural and physical violence) mean that being at COP poses far more significant risks for Indigenous Peoples, ranging from cultural dislocation to murder, and much in between. Moreover, the economic burden, the burden of time, and the burden of proving that Indigenous Peoples are, in fact, the good "stewards" of the environment that they are supposed to be mean that being there and being effective is a costly endeavour.

Conversely, for me the cost of leaving was relatively small. The worst-case scenario was that I would return home and not complete my research. But for the Indigenous people who were there, *not* being at COP would mean not having a say in the future plans for their territories and not having the opportunity to ensure that the violence they have endured is known: they would have no voice. Requiring Indigenous people to venture into the space of the Other in order to affect change reveals a fundamental characteristic of these so-called negotiations. The mere fact that these international discussions that pertain to the communities, territories, and lives of Indigenous Peoples are *always* conducted in non-indigenous places and languages, and according to non-indigenous cultural norms, subordinates indigenous knowledges and ways of being. This in turn selects for the attendance, in most cases, of those Indigenous people and organisations that are most able and most willing to travel to these spaces and to work within the institutional frameworks of them. This process of selection is embedded within and can be understood to be both a cause and a consequence of the ongoing structural violence against Indigenous Peoples. It is rooted in a history of colonialism, of which the United Nations, by implicitly reinforcing its underlying principles, acts as a remnant structure, and it is in this context that Indigenous participation and representation in climate change negotiations is situated.

*

For the remainder of my fieldwork, I spent time in these conferences and in offices and cities, and I felt relatively comfortable doing so. Though the acute experience passed, chronic fear and anxiety remained with me, and I had to find a balance between conducting my research and taking care of my own mental health, and my wife's, as best I could. I navigated my way through my research, negotiating obstacles as they arose. Sometimes I overcame them, and at other times I hid away from them. I barely remember COP20, and some days I did not leave my hotel room at all. But one of the things I do remember happening is my first panic attack, which was triggered when I was driven to the wrong place in a taxi on my way to meeting friends. I suddenly started shouting at the driver to let me out, and I opened the door and rolled out onto the pavement before he had even stopped. I experienced many more over the year that followed, including on the street near the office in Quito (as described in the Chapter 1) and the first time I was back in London after returning to the United Kingdom. It also happened on the Metro in Paris at COP21 when a group of people were chased by armed police onto the carriage I was travelling in. The terror attacks of November 2015 had almost resulted in me cancelling my trip because I was afraid, and when I arrived in Paris, I was drawn into the palpable climate of fear in the city, which left me in a constant state of anxiety.

I missed many research opportunities due to fear, including when I was invited by a member of COICA to visit his community where a REDD+ Indígena Amazónico project was being implemented. Though I came up with what felt like rational reasons at the time, I regretted this decision and many others I made. Knowing that my actions were affected in such a way left me feeling as though I had fallen foul of many of the longstanding criticisms that have been levelled at anthropology and development studies, not least speaking *about* rather than *with* the people I was researching (see, e.g., Sponsel, 1992, pp. 299–301): those whose lives are most directly affected by projects like REDD+. Choosing research sites that felt safe for me means that my research is based upon observations made primarily among only those Indigenous people who could be found in a place that suited *me*, thus illustrating, as well as reinforcing, the asymmetrical power structures within which all research of this kind takes place, and that the research itself contributes to creating. My fieldsites and the people who participated in my research were selected by me, and I was selected by them, and many of my concerns and regrets about this are summarised incisively by Lewis:

> The anthropologist who conducts fieldwork in a colonial setting provides that documentation of differences which functions to support continued subjugation of the group he studies. Secondly, anthropologists promote the exploitation of these differences for their own benefit, both personal and professional. This is demonstrated most blatantly in the attitude of most anthropologists that they have the right to exploit the people they study for their own professional advancement, without having a corresponding sense of commitment to them or their needs. They rarely feel the obligation to "do something" and, in fact, justify their inactivity through recourse to the canon of scientific "objectivity".
>
> *(Lewis, 1973, p. 584)*

In the process of conducting my research and writing it up, I became increasingly aware of its limitations and at the same time stopped trying to overcome them. I came to see myself and my work not as a challenge to extant structures of power, dominance, and colonialism, but instead as being deeply and unchangeably embedded within them. I *knew* all of this even when in the field, but felt unable to do anything about it, and instead sought to protect myself from risk while still gaining as much as I could for myself in the remaining time. I also *knew* how problematic this was, but knowing fear made me *feel* differently about it. I retreated to the safety (and the epistemic failure) of doing what was "sensible" and "rational", and relatively easy. This chapter is intended to shared as openly and honestly as I can why and how this book came to be what it is and to make explicit the problematic characteristics of it. The preceding chapters are based primarily on fieldwork that happened as a result of this, and so should be taken with it in mind.

Notes

1 Named for the oil company, Shell, that founded it in 1937 and abandoned it 11 years later after a series of attacks by indigenous people (Wind, 2017).
2 For discussion of the scientific foundations and evidence for implicit bias, see Greenwald and Krieger (2006).
3 *Los cólonos* in the Amazon region of Ecuador does not always refer to only "white" or even "mestizo" settlers, rather it can mean anyone from outside who has settled in indigenous territories. This potentially includes people from neighbouring communities, or industrial and agricultural workers who might be read as white, Afro-Ecuadorian, mestizo, or indigenous.
4 Communal gardens or small cultivated plots.
5 See previous chapter.
6 *Los Afectados* (the affected) is the term commonly used when referring to 30,000 plaintiffs represented in the lawsuit against Chevron.
7 I returned to visit Pablo once more towards the end of my fieldwork while assisting a documentary film crew in August 2015. We sat in his office in Lago Agrio, near to where I had visited him the first time, and I filmed a conversation between him and a plaintiff in the lawsuit against Chevron. As I listened and watched through the screen on the camera, I heard his story once again. And though I had heard it, read about it and written about it many times over the past six years, and although none of the facts had changed, it became real to me for the first time.
8 I am aware that this might seem like I am ascribing some kind of prophetic status to what the Sápara woman said, or engaging in crypto-Orientalism as Gomez put it (1995, p. 229). I am not. I only mean to say that her idea of what was important was different to mine, and it made more sense to me later. At the time it did not seem important, but I wrote it down.
9 There is some debate around the "White Male Effect" (WME) in studies of risk perception, as significant differences have been observed depending on other characteristics of the society in which studies take place, particularly inequality, leading some to suggest that it is, in fact, better understood as a "societal inequality effect" (Olofsson & Rashid, 2011).
10 For a collection of perspectives on the violence-development relationship, see *Development* (2004) which focuses on violence and includes articles by Escobar, Gibson-Graham, and Sachs, among others. The special issue addresses the question of how "forest dwellers, traditional fisher people, women and men eking out a living in urban slums, tribal and indigenous communities, contract workers, domestic workers far from home, displaced and migrant people, sex workers, orphans and refugees understand development given the violence of their everyday exclusion, exploitation, discrimination and marginalization?" (Kothari & Harcourt, 2004, p. 3).
11 Only a few weeks earlier, Andrés Chota, an Ashéninka activist along with three others (Jorge Ríos Pérez, Leoncio Quincima Meléndez, and Francisco Pinedo), was murdered on September 1, 2014. It was reported that "The circumstances of the deaths are not clear but one local indigenous leader, Robert Guimaraes Vasquez, told a newspaper that illegal loggers bound and shot Chota and companions on the sports field in their village in front of the inhabitants. He said illegal loggers were taking revenge after having been reported to the authorities" (Collyns, 2014). Chota was the leader of the Alto Tamaya-Saweto community and a well-known voice in the struggle for land rights in Peru. He had become particularly visible as a result of recent media attention, including being the

focus of an article in *National Geographic* (Wallace, 2013), which stated that he had received "frequent death threats" from illegal loggers who had been operating within Ashéninka territory.

12 *Cocama* is the word used the Cofán people use to describe non-indigenous Spanish speakers (Cepek, 2018, p. 87).

References

350.org (2014). *Casa Activa*. Retrieved from http://art.350.org/projects/casa-activa/

Adams, A. (1979). An open letter to a young researcher. *African Affairs*, *78*(313), 451–479.

Alexander, J. C. (2004). Toward a theory of cultural trauma. In J. C. Alexander, R. Eyerman, B. Giesen, N. J. Smelser, & P. Sztompka (Eds.), *Cultural trauma and collective identity* (pp. 1–30). London: University of California Press.

Barley, N. (1983). *The innocent anthropologist: Notes from a mud hut*. Long Grove: Waveland Press.

Bauman, Z. (2006). *Liquid fear*. Cambridge: Polity.

Bävman, J (2010). *Stories: Daily Life Along the Pipeline Sote*. Retrieved from http://www.johanbavman.se/?galleries=daily-life-along-the-pipeline-sote

Becker, E. (1975). *The denial of death*. New York: Simon & Schuster.

Behar, R. (1996/2014). *The vulnerable observer: Anthropology that breaks your heart*. Boston: Beacon Press.

Byrne, G. (2014, August 6). Going out of my mind. *Sociology Lens*. Retrieved from https://www.sociologylens.net/topics/culture/going-out-of-my-mind-in-jandiayacu/13520

Capella, M., Jadhav, S. S., & Moncrieff, J. (2020). History, violence and collective memory: Implications for mental health in Ecuador. *Transcultural Psychiatry*, *57*(1), 32–43.

Cepek, M. L. (2018). *Life in oil: Cofán survival in the petroleum fields of Amazonia*. Austin: University of Texas Press.

Collyns, D. (2014, September 9). Illegal loggers blamed for murder of Peru forest campaigner. *The Guardian*. Retrieved from https://www.theguardian.com/world/2014/sep/09/illegal-loggers-blamed-for-of-peru-forest-campaigner

Darrell, A., & Pyszczynski, T. (2016). Terror management theory: Exploring the role of death in life. In G. S. Nisbett & L. A. Harvell (Eds.), *Denying death* (pp. 1–15). London: Routledge.

de Becker, G. (1997). *The gift of fear*. London: Bloomsbury.

Du Bois, W. E. B. (1903/2008). *The Souls of Black Folk*. Oxford: Oxford University Press.

Dunlap, A. (2017). Wind energy: Toward a "Sustainable Violence" in Oaxaca. *NACLA Report on the Americas*, *49*(4), 483–488.

Fabian, J. (1983/2014). *Time and the other: How anthropology makes its object*. New York: Columbia University Press.

Finucane, M. L., Slovic, P., Mertz, C. K., Flynn, J., & Satterfield, T. A. (2000). Gender, race, and perceived risk: The "white male" effect. *Health, Risk & Society*, *2*(2), 159–172.

Gomez, L. O. (1995). Oriental wisdom and the cure of souls: Jung and the Indian East. In D. S. Lopez (Ed.), *Curators of the Buddha: The study of Buddhism under colonialism* (pp. 197–250). London: University of Chicago Press.

Greenberg, J., Pyszczynski, T., & Solomon, S. (2015). Thirty years of terror management theory: From Genesis to revelation. *Advances in Experimental Social Psychology*, *52*, 1–70.

Greenwald, A. G., & Krieger, L. H. (2006). Implicit bias: Scientific foundations. *California Law Review*, *94*(4), 945–967.

Haraway, D. (1988). Situated knowledges: The science question in feminism and the privilege of partial perspective. *Feminist Studies*, *14*(3), 575–599.

Hegeman, E., & Stocker, M. (1996). *Valuing emotions*. Cambridge: Cambridge University Press.

Hopfgartner, K., Moreano, M., & Santillana, A. (2016). La Política Económica de la Reforma a la Gobernanza del Agua y las Implicaciones para la Desigualdad Territorial El Caso de Ecuador. *RIMISP: Santiago de Chile, Chile*. Retrieved from https://www.rimisp.org/wp-content/files_mf/1470164857 Lapoliticaeconomicadelareformaalagobernanzadelaguaylasimplicacionesparaladesigualdadterritorial.pdf

Jamison, K. R. (1995). An Unquiet Mind: A Memoir of Moods and Madness. New York: Vintage Books.

Janack, M. (2002). Dilemmas of objectivity. *Social Epistemology*, *16*(3), 267–281.

Kothari, S., & Harcourt, W. (2004). Introduction: The violence of development. *Development*, *47*(1), 3–7.

Lewis, D. (1973). Anthropology and colonialism. *Current Anthropology*, *14*(5), 581–602.

Macias, T. (2016). Environmental risk perception among race and ethnic groups in the United States. *Ethnicities*, *16*(1), 111–129.

Ofrias, L. (2018, March 20). Contamination of the social fabric: Sápara leaders' resistance to oil companies in the Ecuadorian Amazon and the narrow vision of environmental impact studies. *Engagement*. Retrieved from https://aesengagement.wordpress.com/2018/03/20/contamination-of-the-social-fabric-sapara-leaders-resistance-to-oil-companies-in-the-ecuadorian-amazon-and-the-narrow-vision-of-environmental-impact-studies/

Olofsson, A., & Rashid, S. (2011). The white (Male) effect and risk perception: Can equality make a difference? *Risk Analysis: An International Journal*, *31*(6), 1016–1032.

Páz y Miño, P. (2008). *Amazon Watch Blasts Consultant Sam Singer for Designing Chevron Attack on Goldman Prize Winners*. Retrieved from https://amazonwatch.org/news/2008/0416-amazon-watch-blasts-consultant-sam-singer-for-designing-chevron-attack-on-goldman-prize-winners

Perreault, J. M. (1985). *Writing selves: Contemporary feminist autography*. Minneapolis: University of Minnesota Press.

Reed, P. (2011). REDD+ and the indigenous question: A case study from Ecuador. *Forests*, *2*(2), 525–549.

Rich, A. (1984). Notes toward a politics of location. In A. Rich (Ed.), (1994). *Blood, bread, and poetry: Selected prose 1979–1985*. New York: Norton.

Smith, L. T. (1999). *Decolonizing methodologies: Research and Indigenous peoples*. London: Zed Books.

Smolewski, M., & Wesley-Esquimaux, C. C. (2004). *Historic trauma and aboriginal healing*. Ottawa: The Aboriginal Healing Foundation.

Solíz, M. F. (coord.) (2017). *Salud Colectiva y daño psicosocial en las familias de la comunidad de Tsuntsuim*. Quito, Acción Ecológica, Alames Ecuador, Conaie-Confeniae, Movimiento por la Salud de los Pueblos, Luna Creciente.

Solomon, S. (2015). Interview by N. Warburton. Sheldon Solomon on fear of death. *Social Science Bites*. Sage. Retrieved from http://www.socialsciencespace.com/wp-content/uploads/Sheldon-Solomon-trans.pdf

Sponsel, L. E. (1992). Information asymmetry and the democratization of anthropology. *Human Organization*, *51*(3), 299–301.

Taussig, M. (1984). Culture of terror – Space of death. Roger Casement's Putumayo report and the explanation of torture. *Comparative Studies in Society and History*, *26*(3), 467–497.

Taussig, M. (2003). *Law in a lawless land: Diary of a Limpieza in Colombia*. London: University of Chicago Press.

Tegantai (2018, April 27). Presidenta de Nacionalidad Sápara Recibe Amenaza de Muerte. Retrieved from http://www.ecuadorinmediato.com/index.php?module=Noticias&func=news_user_view&id=2818836266

Tsing, A. L. (2015). *The mushroom at the end of the world: On the possibility of life in capitalist ruins*. Princeton: Princeton University Press.

Wallace, S. (2013). Mahogany's last stand. *National Geographic*, *April 2013* [Online]. Retrieved from https://www.nationalgeographic.com/magazine/2013/04/mahogany-last-stand/

Wind, D. D. (2017, December 19). What faith built in a place called 'Shell'. *Huffington Post*. Retrieved from https://www.huffingtonpost.com/dorian-de-wind/what-faith-built-in-a-pla_b_13714832.html?guccounter=1

CONCLUSIONS

Re-Membering Research

In writing this ethnographic account of my experiences and observations during fieldwork, my hope is that it will allow the reader to reach some of their own conclusions regarding the questions posed in the methodology. Some direct responses to these have been offered in the text, while others have been alluded to or inferred in the stories that have been told. I imagine, however, there are also many other ways that I have not considered in which the information presented could be interpreted. This is, after all, a partial re-membered (see Haraway, 1988; Smith, 1999, p. 147) account of almost five years of research in a diverse set of fieldsites in three countries, and it presents only an overview of key observations from numerous semi-structured and unstructured interviews with both Indigenous and non-indigenous people. Through the reflections in this book, I have put back together a disjointed and chaotic personal experience with the intention of providing a coherent and comprehensible narrative that is intended to communicate how I reached the views I have expressed.

As the encounters and experiences described have shown, my research took a very different path to what I originally intended, and it has required a significant amount of reflection and reconsideration of my approach, as well as the initial research questions. Here in the conclusions, I summarise my thoughts on two themes that emerged through this process. First, I consider the empirical data and focus on the relationship between indigeneity and climate change, and I then reflect on the methodological implications has for my work (and research more generally) when attempting to answer questions relating to the experiences of marginalised people. In the final section of this chapter, I provide a tentative response to one of my main research questions: what, if anything, might Indigenous Peoples "know" about climate change that "we" (in the so-called developed nations, as Manari puts it) do not? And yet, while I do offer my opinion, I make my argument with a paradox in mind: part of what I am attempting to convey is that research and academic writing are fundamentally limited in their capacity to understand and communicate particular forms of knowledge that are created in other ways.

Nonetheless, this book is the tool I have available to me here and now. As such, if I am going to make an academic contribution, it is to suggest that – from my perspective (as a white, British man situated within the epistemological community of Western academia) – it appears that what it means to be indigenous in relation to climate change is continuously being made up through a process of selection. The category comes into existence as a result of interactions between a wide-ranging assortment of actors and institutions, through which particular characteristics and ways of being are selected and deselected based upon how well suited they are to the places in which they emerge. This process is subject to a number of structural limitations that mean that, although all actors can be said to have some degree of agency in each interaction, the eventual outcome (generally speaking) is not equal or equitable. This is because these negotiations begin from a place where power is inequitably distributed, and this asymmetry is rooted in a history of social, cultural, and epistemological dominance. Though in a sense it is a relic of a violent colonial past, this injustice is situated very much in the present. With that in mind, it is important to point out that in no uncertain terms, the book I have written is both subject to and implicated in the very same problem of selection I have been illustrating.

Making Up an Optimal Other?

The first two ethnographic chapters of this book "Negotiating Indigeneity at the Indigenous Pavilion" and "Encountering the Optimal Other" provided snapshots of my observations while at the COP events in Lima and Paris, and while with COICA in Quito. For me, these places represented a retreat to a familiar space where I felt relatively safe and where I was able to spend time with a group of Indigenous people who were accommodating and open to my presence in their spaces. The middle part of my research, the COICA office in Quito, was practically the most straightforward, but it was particularly challenging because it required me to make a significant adjustment to my preconceptions of what it means to do research with Indigenous Peoples, as well as what it means to *be* indigenous. Much of the time I felt as if I was not really doing anthropology, or any research, at all. Sometimes it was mundane and boring because the office and the people were, in many respects, like those in any major city of the world. Of all the places I visited, though, it also most resembled the middle ground described by Conklin and Graham (1995): it was certainly mutually comprehensible, to the point where it was difficult to say, from my perspective, what was uniquely indigenous about it at all. Many of the people I spent time with there were Indigenous, but rarely would I see them being *Indigenous*.

Having encountered members of COICA at COP20 in Lima some months earlier, I had at first seen some of them in a very different light. They had presented themselves and the people they represent at the Indigenous Pavilion in a way that was much more closely aligned with the "green development fantasies" (Tsing, 2007) found in Western minds like mine. This was apparent in their bodily presentations, their rhetorical use of the language of "stewardship" and "guardianship", and in the physical manifestation of an indigenous space in a non-indigenous place: a bronze statue of a tribe and a sanitised simulacrum of a rainforest encased in the white, ultra-modern, geodesic dome. The office in Quito, however, seemed to be a space positioned between this highly curated image of indigeneity and the "real" Indians described in Chapter 5, and it is this that led me to begin to draw a distinction between being indigenous and being *Indigenous*. The former can perhaps be best understood as a biosocial fact (see Hartigan, 2013, p. 4): people and things can be described as being indigenous *to* somewhere by virtue of cultural or biological lineage, or by embodying the cultural practices, norms, and markers that are associated with specific geographic regions. But to be *Indigenous*, not to anywhere in particular but instead as part of a category that is distinct from being non-indigenous, is something altogether different. To be *Indigenous* appeared to me to be an active state, which is constantly changing and being renegotiated. It is a transient identity that is shaped by and for the space in which it emerges and by the purpose for which it is being used, but its impermanence and inconstancy make it no less real. Similarly, for me, it is situated in the "then and there" because I am writing about it and reconstructing it from how I remember it. But as an identity being *Indigenous* is not a fading remnant of the past that needs to be "preserved" and it is instead a contemporary political reality that is present and is *in* the present.

I occasionally glimpsed echoes of this particular form of indigenous identity emerging within the cultural space of COICA, particularly at the 30th Anniversary events where the dancers and the beauty pageant contestants were brought in to perform. The indigeneity they represented did not have to be indigenous *to* anywhere in particular, but instead, it signified a shared pan-indigenous identity, the important part of which was that it was clearly *not* non-indigenous. My initial response to this was a feeling of unease and cynicism: I questioned the authenticity of such displays and began to consider what it said about COICA's legitimacy. There are certainly questions to be asked regarding how an organisation like COICA, which is culturally and geographically distant from any indigenous territory, can legitimately claim to represent hundreds of thousands of Indigenous Peoples. But as time went on my scepticism of the organisation lessened, giving way to a question that echoes one of the slogans used by COICA in the build-up to COP21: "If not us then who?". At the time, I took this question to mean if they (Indigenous Peoples) do not take a stand to protect the rainforest territories in which their communities live, then who will? But one could equally ask who, if not an organisation like COICA, by speaking in the language of the dangerous Other world, and by "identifying with all that is violent and morally condemned" (Cepek, 2012, p. 90), can ensure that Indigenous Peoples of the Amazon have a voice in these negotiations at all?

I am aware that the comparisons made to shamanism risk further romanticising or essentialising Indigenous Peoples, but that is not my intention. Instead, by making this connection, I seek to highlight a different problem: to be taken seriously and to be seen as "legitimate" representatives, authenticity is demonstrated in exaggerated bodily presentations and in perpetuating specific "positive" stereotypes that are associated with being indigenous. This represents a kind of trap, what Brock, Dove, Hebdon, and Ludlow (2016) called a "double bind", whereby if a person does not appear sufficiently different, they cannot claim to represent the Other, but if they try too hard, they are seen as "inauthentic". At the same time, to be heard and understood Indigenous people speak in non-indigenous languages (mostly Spanish in the case of COICA, which itself causes some internal issues within the organisation), and they express ideas and evidence using distinctly "Western" concepts such as "scientific" data, tCO_2e, and the notion of environmental stewardship. For example, in the indigenous alternative to REDD+ presented by COICA, their "holistic" cosmovisions for their territories are reduced to 24 functions of the rainforest, and they demonstrate the effectiveness of their stewardship through carbon mapping. This is why, on the final day of COICA's 30th Anniversary conference in the room of the university in Quito, I was struck by the symbolism of Perugachi's painting of "Liberty in the Andes": the event and the painting both illustrated the belief that Indigenous Peoples can be emancipated from a colonial history of violence and oppression *by* the ideology and institutions of Western liberty, if they can only hold on tight enough.

This became most visible to me at the Indigenous Pavilion during COP21 in Paris. It was intended to be a united platform from which Indigenous Peoples from around the world can be included in the UNFCCC negotiations, and to have secured such a prestigious position in a space like this is undoubtedly a huge success. To have no voice at all would likely be a tragedy for Indigenous Peoples. And yet, the very presence of such "representatives" at COP (and at other events where projects like REDD+ and PES are being discussed, such as the workshop with the GOE and the UN discussed in Chapter 4) lends a degree of legitimacy to the events themselves, and to the paradigm of neoliberal "sustainable development" that they extol. But these interactions are far from equal and, I believe, far from inclusive of Indigenous Peoples' perspectives. Notions of holistic management of forests, indigenous cosmovisions, or even non-market-based approaches, are taken seriously *only* if they can be shown to be economically viable and environmentally effective in terms of reducing (or offsetting) carbon emissions. While governments, IGOs, and "Big Green" NGOs (all of whom COICA works with) tend to speak in terms of "opportunities" and "co-benefits" of "climate change mitigation" and "sustainable development", it is the territories, resources, cultural survival, and the *lives* of Indigenous Peoples that are actually being negotiated, not abstract tonnes of carbon dioxide equivalents.

Thus, to be "given" a voice in these negotiations, Indigenous Peoples demonstrate that they are, in fact, the guardians of the rainforests and the so-called stewards of our sustainable development fantasies. But they must also demonstrate a willingness and capability to serve this function within the dominant ideological and epistemological framework. Such relational encounters between epistemological communities shape and are shaped by systemic asymmetries of power, and this leads to the selection of an Optimal Other: still savage, but not *too* savage; ecologically noble, but also still sympathetic to the concept of "development". Those who walk this tight rope effectively are permitted to speak, while those who choose not to (or do so ineffectively) are not. This does not just mean that some Indigenous people are selected and others are deselected, rather it means that through a process of selection the very notion of what it means to be *Indigenous* is reified. Those who do not identify with this emergent category are invisiblised, and silenced, and their identities are entirely removed from what it means to be *Indigenous*, particularly in relation to climate change but also in Western minds more generally.

For the Indigenous people who are being *Indigenous* at COP, their presence also serves a different purpose: it allows for aspects of indigenous cultures that overlap with Western fantasies and the myth of altruistic sustainable-developmentalism (or even arguments *for* colonialism) to become a political tool in the form of symbolic indigenous capital. It is used to demonstrate that Indigenous Peoples are not simply passive beneficiaries of sustainable development projects, nor are they destructive consumers of the forests in which they live, but instead they have knowledge and experiences that give them a unique ability to care for the natural world and, therefore, contribute to mitigating climate change. This ecological nobility is euphemised to "stewardship", and this function is an explicit precondition of indigenous legitimacy within the UN framework: Indigenous Peoples are granted rights, and the right to speak, not because they are *People*, but because they are *Indigenous*. They are seen as *useful*.

Viatori noted that the Sápara, for example, do this with great skill. They tap into discourses relating to environmentalism and frame their cultural practices within the parameters of essentialism when (re)presenting themselves in order to evoke "authenticity". It is done knowingly and effectively (Viatori, 2007, p. 112). My overall impression of spending time with Indigenous people was that as individuals they may or may not be ecologically noble, but they are almost always ecologically savvy, and are well versed in the power of the discourse. This can lead to a greater degree of interest and investment in Indigenous Peoples' organisations and to an increasingly legitimised role in the political world at multiple levels. But, as Bourdieu noted, the contradictory nature of such symbolic capital means that although it can create symbolic power, it tends to legitimise those who are already most assimilated into the institutional structure:

> ... it is the power granted to those who have obtained sufficient recognition to be in a position to impose recognition. In this way, the power of constitution, a power to make a new group, through mobilization, or to make it exist by proxy, by speaking on its behalf as an authorized spokesperson, can be obtained only as the outcome of a long process of institutionalization, at the end of which a representative is instituted, who receives from the group the power to make the group.
>
> *(Bourdieu, 1989, p. 23)*

Beyond this dual-legitimising role, the presence of some Indigenous people at COP has another more complex and, I think, less intentional effect. People who visit the Pavilion with the explicit intention of seeing Indigenous Peoples, including activists, government officials, journalists and reporters, and academics (like me), do so in a manner that replicates a process (re)discovery. The allure of such spaces is that these "exotic", "primitive", "spiritual", "noble" Peoples and their worlds can be encountered without a white man ever having to leave the comfort of the conference. In this sense, the physical presence of the Indigenous Other fulfils the role described by Viveiros de Castro: giving meaning to the non-indigenous Self, a world without which is "unthinkable" (Viveiros de Castro, 2011, p. 73). Thus, a shared indigenous identity that bonds millions of people from thousands of tribes, nations, and peoples around the world in a single, supposedly homogenous, political category emerges. And, meanwhile, it satiates the colonial desire to discover, explore, and include, but also subjugate, Other people and Other ways of being and knowing.

But of course, nobody is being "discovered" in any real sense. The discourse of discovery is itself rooted in a Eurocentric epistemological position that treats Western knowledge not simply as *superior* to indigenous knowledges, but as the *only* legitimate way of knowing, as if the white eye sees from the centre (Rich, 1984). Said (1978) described how through "discovery", observation and description, the Other is *almost* invented in, by, and for the Western mind. The simple form of this argument when applied to Indigenous Peoples could be to say that they were never discovered because *they* already knew that *they* existed,[1] but this is an incomplete description. As Spivak noted in the 1993 essay *Can the Subaltern Speak?* knowledge is never objective and it is never innocent, it is constructed by and for the purposes of those who create it and the "remotely orchestrated, far-flung, and heterogeneous project to constitute the colonial subject as Other" is "The clearest available example of such epistemic violence" (Spivak, 2013, p. 76). Hacking's theory of how categories of people are "made up" by the very act of categorisation (Hacking, 1986, p. 226) complements this. In this sense, Indigenous people could never be discovered because they did not exist at all. Instead, they were made up in the minds of European colonisers and have since become a new category of people who continue to change and be (re)created, made up again and again through encounters, negotiation, and conflict. As Fanon put it:

I came to this realistion during my research, but it isn't new. Frantz Fanon, for example, made the same observation in 1961 in *The Wretched of the Earth*:

> The settler and the native are old acquaintances. In fact, the settler is right when he speaks of knowing "them" well. For it is the settler who has brought the native into existence and who perpetuates his existence.
>
> *(Fanon, 1963/2004, p. 2)*

In reference to the Indigenous Pavilion, then, the place itself can be viewed as a simulacrum: a physical embodiment of the invented Other world with no origin or underlying reality. But the simulacrum, as Baudrillard (2004, p. 169) put it, is "true": *Indigenous* is indigenous, there is no other "real" indigeneity that underpins this place or this shared identity. Instead, the Pavilion is a focal point from which knowledge (which is neither objective nor innocent) of what it means to be *Indigenous*, particularly in relation to climate change, is (re)created and (re)negotiated, and it is then leveraged for particular political ends. This way of being *Indigenous* then circulates through to every extremity of the social body, influencing what it means to be indigenous in every part, and at every level, of society. And by (re)constituting indigenous as Other, it can be understood as indicative of ongoing structural and epistemic violence, which recreates and even justifies the asymmetries and inequalities of colonialism. So, in places and spaces where responses to climate change are being negotiated, notions of sustainability and indigeneity have become intertwined. People who were once treated as less than human became human but "savage", then "noble", and eventually "ecologically noble"; the "beneficiaries" of development became "stakeholders", and then "actors", and they are now imagined to be "stewards" or "guardians" of the planet. Today, Indigenous Peoples are presented as the leaders in the fight against climate change, "they" are seen as capable of protecting and saving the environment that "we" have destroyed, and "they" are at times positioned as morally or ethically (or at least ecologically) superior to "us". But even through this political transformation in the Western mind, Indigenous Peoples have always remained Other. To return to Spivak's observation, this requirement – for Indigenous Peoples, if they are to be "given voice" in non-indigenous places, to *prove* that they are indeed recognisable examples of the colonial subjects who are, by definition, constituted as Other – is a clear example of epistemic violence (Spivak, 2013, p. 76).

Epistemic Failings of the Colonised Mind

The earlier part of my fieldwork, which is covered in Chapters 5 and 6 ("Failures and Frustration during Research Encounters", and "Fear, the Real, and the Other"), was where I began to reconsider my own position and the role of conducting research that pertains to Indigenous Peoples. There are two main reasons for dedicating so much time to discussing what could, in some sense, be considered "failed" research rather than condemning it to my private fieldnotes (Wolcott, 2005, p. 214). First, I believe there are some lessons that can be learnt from the process I went through, whereby I attempted to come to terms with and overcome my own identity and position within extant structures of colonialism (which I essentially failed to do). Second, certain parts of this experience offer illustrative examples of the process of selection that occurs when working with Indigenous people: those who are the "easiest" to work with - the ones who offer their time, their energy, their knowledge, and their culture to outsiders, often for seemingly little or nothing in return - are the ones whose voices are most likely to be reproduced and amplified. At these interfaces, the process of selection becomes an expression of power, but also a means of resistance. "We" select (or deselect) "them", but "they" simultaneously do the same to "us".

From the beginning, my two primary collaborators acted as gatekeepers to the Sápara community, and my relationship with them was complicated. Not for the first time, but in a far more experiential sense than ever before, I became acutely aware of my embodied identity as a white, thirty-something, British, male, researcher and began to question what this might mean to the people I was encountering. Each of these characteristics positioned me in contrast to the two Indigenous women with whom I was working, and it was here that the categories of "I" and "Other" began to take on much more meaning for me: *my* identity as non-indigenous was reified. An example of this was when I was given the task of choosing a representative of the Sápara people despite having only just met them and knowing very little about them. I could see no real reason why I had been granted this power, but when I decided to not make the decision and instead *give them* the freedom to choose for themselves, *it was still my choice* to do so. Conversely, when Maria decided she no longer wanted to work with me, my access to the Sápara community was unilaterally revoked and there was nothing I could do about it without potentially causing considerable harm. The difference in relation to my research, though, was that I was still able to continue it by seeking out Indigenous people with whom I could spend time, and whose presence in my research lends it a degree of apparent legitimacy, despite its problematic nature. As noted, this pattern of interaction is repeated at multiple levels of society, from the individual and interpersonal to the political and international, not just in academic research.

The main methodological concern that arose during my project is how, and indeed if, it is possible for a person like me to conduct research that pertains to people who (in relation to Western academia, at least) are marginalised without contributing to the very epistemic violence that I have observed and critiqued. I don't have a good answer to this. I have attempted to avoid doing harm and have been explicit about my positionality, ensuring that any assertions are made with some degree of epistemic humility. In writing, I chose to do so in reverse, beginning where my research ended and from the perspective I developed during writing and then working my way back through the experiences and encounters that brought me to this position. My lived experience of interpersonal relationships, of violence and of fear, and of privilege and freedom became part of my analysis, which I have included in this book with the intention of creating space for the lived experiences and embodied knowledges of others to be taken seriously, rather than treating them as metaphors or trying to "translate" them. I also wanted to avoid writing about Others while acting as if I were invisible or neutral. I worked through these experiences by engaging in a self-critical process of reflection (Ploder & Stadlbauer, 2016, p. 754) and tried my best to do my "homework" (Spivak, 1990, pp. 62–63), neither of which should be understood as having made my observations any more objective. Deciding to be open and honest about my experiences, including those that were traumatic and those that felt like "failures", is only intended to subvert, and go some way toward transgressing, the apparent tendency of researchers (particularly at the level of writing a doctoral thesis, but also when attempting to publish) to *write selectively* in order to impress the reviewer or reader with their skills and knowledge, rather than *say what actually happened.*

This tendency, I believe, is at least partly a response to the pressure to prove oneself within established hierarchies of knowledge, which in turn distorts the epistemological world of academia. I have attempted to be honest about and critical of my own authority and the asymmetric power relationship between me, my research, and the subjects of it. To do so, I employed a number of reflexive tools, including writing about myself in first and third person when recounting my observations depending on whether I considered myself to be more "inside" or "outside" of what was taking place. I adopted a free voice when it allowed for a more effective and affective presentation of my thoughts, particularly when discussing personal experiences and emotions. The choice to include selected autoethnographic narratives was intended to soften the boundary between the researcher and the research and to make clear that the book I have written is subject to and implicated in the same problems of biases and selection that I have discussed throughout. My work is, of course, "contaminated" (Tsing, 2015) by this; it is neither "objective" nor "rational": it is a highly subjective account of a partial perspective.

Thinking about Knowing and Feeling: Indigenous Peoples and Climate Change

In considering the ways that fieldwork can cause harm, in the final part of this book I drew the narrative into an explicitly subjective account, and I used fear as lens through which to consider my experience and the experiences of Others. It represents a moment in time when I was forced to reconsider much of what I thought I knew and to begin seeing abstract ideas, including "the field", the people I encountered in it, and their experiences and feelings, as real. Of course, I cannot expect the reader to feel what I felt, and as such it follows (if feeling and knowing are as connected as I now believe they are) that I cannot perfectly communicate what I know. The same can be said for making assertions regarding what Indigenous

Peoples know about climate change that "we" do not: such claims risk falling foul of many of the criticisms I have made so far. Nonetheless, I will allow myself to engage in some conjecture and *say what I think and feel about this question*. I do happen to think, now, that there is something unique about Indigenous Peoples' perspectives on climate change, but I didn't reach this view for the usual reasons. It is not simply that by virtue of being indigenous *to* somewhere, some people necessarily have more or better knowledge about how to protect rainforests: for me, this argument veers too close to the myth of the ecologically noble savage. There is undoubtedly a great deal of unique and "valuable" knowledge in indigenous communities (including situated, experiential, and embodied knowledges), but there is also a vast amount of important and very useful information about climate change that has emerged through scientific investigations and other studies that originate in Western academia. Knowledge alone is not enough to explain the differences between indigenous and non-indigenous perspectives. What I observed during my fieldwork is that among many Indigenous people, climate change and environmental concerns are *perceived*, differently. Indigenous Peoples seem to know and feel that these risks are *real*.

The experiences of an individual or a group, particularly experiences of violence or trauma, can and do change the ways in which risk is perceived. Fear and other emotions, though often treated as irrational, do not simply degrade the "objective" viewpoint but instead lead to the emergence of a different partial perspective that is not inferior and, in some instances, can be *more* rational than artificial objectivity. The historic violence against Indigenous Peoples (in Ecuador, but also around the world) can potentially lead to the emergence of embodied knowledges, which in some instances are influenced (or contaminated) by through the experience of collective trauma. I have known and felt fear myself, and I have seen it among Indigenous Peoples, and in perceiving risk more acutely I believe that, because of this, they see, understand, and *know* climate change (among other risks) more intuitively. Conversely, the epistemological community of the Western world seems to hold an upside-down view of knowledge: particular ways of knowing are considered to be necessarily more accurate (and therefore better) *because* they are abstracted from, and therefore supposedly immune to, the inaccuracies that come with subjectivity. But the subjective experience is arguably what gives knowledge meaning: with no intuitive, embodied perception of the risks of climate change, with no *fear*, there may be no impetus for action; no *hope*.

Taking risks with fear in this way might be related to the characteristic "principle of hope" that is so often attributed to the political activity and movements led by Indigenous (as well as other subaltern) Peoples

(Cusicanqui, 2012, p. 96). It has long been a "characteristic of the contemporary indigenous politics which is often criticized by non-indigenous scholars, because it is viewed as being overly idealistic" (Smith, 1999, p. 91). But it is unclear at this point how, without an "overly idealistic" movement, climate change can be effectively addressed. It is important also to note that such hopefulness can be inspired by the lived experience of, and the fear of, a real imminent danger:

> People who can be afraid see the world in terms of possibilities, in terms of things getting better. Thus, they are hopeful. They see the world from the standpoint of an active person, rather than, say, a passive, or disinterested, or played out spectator. They see the world in terms of what they can do.
>
> *(Hegeman & Stocker, 1996, pp. 255–256)*

If there is one observation that sums up my research and what I am trying to communicate in this book it is this: the paradoxical outcome of this line of thought is that the person in an apparent position of relative power, and who is privileged enough to not feel fear (and can therefore be "rational", "objective", and "reasonable"), might be trapped in an epistemic failure whereby they acknowledge that climate change is a catastrophe but, because it does not feel *real*, they are left powerless or utterly unmotivated to do anything about it. Because they are fearless, they are hopeless. Any action that is taken is therefore restricted by ideas of what is *impossible*, rather than being motivated by what is *possible*: both the potential catastrophe and taking serious action to avoid or mitigate it are unimaginable when one feels insulated from risk while also being constrained by the "conventional wisdom" of what is realistically achievable.

A pertinent example of this is the dominance of market-based approaches to mitigating climate change. Almost nobody, even the institutions that propagate such notions (like the UN), really seems to think that such mechanisms will be effective enough to halt or reverse environmental damage. Instead, they are viewed as reasonable, rational, and sensible efforts to fix an abstract problem that is not taken seriously: they are based in neither fear nor hope. And yet, to weight "costs and benefits" (to use the development lingo) so heavily in the present or in the immediate future is entirely *irrational*, as is prioritising the short-term economy over the future of the environment. If it is true that the "White Male Effect" (Finucane, Slovic, Mertz, Flynn, & Satterfield, 2000) leads to extremely low levels of risk perception, or even if it is a "societal inequality effect" (Olofsson & Rashid, 2011), then it is a real and imminent risk to the rest of society that "we" are the ones who hold the most decision-making power in negotiations about climate change, while those who perceive risks more clearly and who are most motivated to act are also the most marginalised.

No Centre from Which to See

This book represents the end of a personal cycle of learning, which amounts to over 15 years spent in academic institutions studying development, international relations, and anthropology. Unfortunately, some things have not changed a great deal during this time. I still consider academia to be a tool of social control. Through it, categories of people are made up and subsequently Othered, or are assimilated into particular epistemic communities, and through these epistemological hierarchies are maintained. But I also see it as having the potential to offer individual social mobility and to be a space within which these hierarchies can be challenged and subverted, provided research is approached with an emancipatory imperative. In the years since I conducted the bulk of the research for this book, there have been some notable shifts in this direction, not least the mainstreaming (to some extent) of decoloniality and anti-racism in UK academic institutions. This, however, has left open questions regarding the possibilities for epistemic and racial justice emerging from within capitalist institutions that are infused with the logics of coloniality (Hall, Ansley, & Connolly, 2023). With this problem in mind, I included in the introduction to this book a quote that I have tried to keep in focus while writing. I put it on a post-it note and stuck it on my computer screen, and I think it is worth repeating it here:

> I do not any longer believe – my feelings do not any longer allow me to believe – that the white eye sees from the centre. Yet I often find myself thinking as if I still believed that were true. Or rather, my thinking stands still. I feel in a state of arrest, as if my brain and heart were refusing to speak to each other.
>
> *(Rich, 1984, p. 226)*

At first, I took this to mean that to see from the centre my positionality needed to be acknowledged and overcome. Now, however, I understand that: the white eye does not see from the centre because there is no centre from which to see. There is no rational, objective viewpoint, only myriad partial perspectives, all of which are subjective and all of which are coloured by experience and emotion. This acknowledgement has allowed my brain and heart to speak to each other, which has helped me to move beyond the state of arrest I felt for such a long time after I started to feel that the abstracted things and people I had been writing and thinking about were real.

If I were to embark upon this project again from the beginning, knowing and feeling what I do now, I would do so very differently. I might not even do it at all. I certainly would not begin from a position of "curiosity", seeking to extract knowledge from Indigenous Peoples and find out what "they" know about climate change or what it means to be "indigenous". I am more sceptical now than ever about the place of anthropology and development studies in speaking about themes that relate to unequal distributions of power because the disciplines are themselves implicated not only in the construction of the indigenous/non-indigenous dichotomy but also in (re)creating and sustaining the asymmetries between them. I might instead start from the question of what it means to be *non*-indigenous, and I might consider how this supposedly "neutral" position limits the effectiveness of climate change mitigation policies (like REDD+) that emanate from the remnant structures of colonialism (like the UN).

I consider feminist and decolonial approaches to research to offer many effective ways to begin overcoming this problem, and through them I have endeavoured to unlearn some of what I thought I knew and to take the views of Others seriously. But I cannot claim to have conducted decolonial or feminist research because I did not begin to engage with these methods in any serious way until well into my fieldwork, and even then, I did so gradually and inconsistently. By that point, the pressure of time coupled with my own desire to be in places and spaces where I felt safe, meant that the Indigenous people with whom I spent most time (and who are therefore most prominent in this book) were mostly a small elite group within a much larger marginalised group. Though this made for some interesting observations, my concern is that I may have contributed to two contradictory and problematic discourses. The first is the idea that these particular Indigenous people *are* "legitimate" representatives of all people who might identify themselves as Indigenous, and the second is that my critical perspective could be interpreted as suggesting that they are somehow "inauthentic" and, therefore, are *not* legitimate. Either or both of these arguments could be used to undermine Indigenous Peoples' struggle for representation in global institutions or to de-legitimise their voices in climate change negotiations. I have intended, though, to illustrate that being *Indigenous* performs a complex role in such spaces, and taking on this identity is neither more nor less "authentic". Rather, it represents a particular form of indigeneity that has been made up by and for the context in which it has emerged, and although it may at times resemble the Indians of Western green development fantasies, it has also facilitated the construction of a category of people who are among the most successful and powerful advocates for Amazonian Indigenous Peoples at the international level.

Note

1 Smith describes the chronology of the "cultural decay" that followed this "discovery" as being generally viewed by the colonizer as a progression through "(1) initial discovery and contact, (2) population decline, (3) acculturation, (4) assimilation, (5) 'reinvention' as a hybrid ethnic culture", while indigenous perspectives would be more likely to describe these phases as "(1) contact and invasion, (2) genocide and destruction, (3) resistance and survival (4) recovery as indigenous peoples" (Smith, 1999, p. 91).

References

Baker, L., Brock, S., Dove, M. R., Hebdon, C., & Ludlow, F. (2016). The double binds of indigeneity and indigenous resistance. *Humanities*, *5*(3), 53.

Baudrillard, J. (2004). Simulacra and simulations. In M. Poster (Ed.), *Selected writings* (pp. 169–187). Cambridge: Polity (original work published 1981).

Bourdieu, P. (1989). Social space and symbolic power. *Sociological Theory*, *7*(1), 14–25.

Cepek, M. (2012). *A future for Amazonia: Randy Borman and Cofán environmental politics*. Austin: University of Texas Press.

Conklin, B. A., & Graham, L. R. (1995). The shifting middle ground: Amazonian Indians and eco-politics. *American Anthropologist*, *97*(4), 695–710.

Cusicanqui, S. R. (2012). Ch'ixinakax utxiwa: A reflection on the practices and discourses of decolonization. *South Atlantic Quarterly*, *111*(1), 95–109.

Fanon, F. (2004). *The wretched of the earth* (R. Philcox, Trans.). New York: Grove Press (original work published 1963).

Finucane, M. L., Slovic, P., Mertz, C. K., Flynn, J., & Satterfield, T. A. (2000). Gender, race, and perceived risk: The "white male" effect. *Health, Risk & Society*, 2(2), 159–172.

Hacking, I. (1986). Making up people. In D. E. Wellbery, M. Sosna, & T. C. Heller (Eds.), *Reconstructing individualism* (pp. 222–236). Stanford: Stanford University Press.

Hall, R., Ansley, L., & Connolly, P. (2023). Decolonising or anti-racism? Exploring the limits of possibility in higher education. *Teaching in Higher Education*, pp.1–17.

Haraway, D. (1988). Situated knowledges: The science question in feminism and the privilege of partial perspective. *Feminist Studies*, *14*(3), 575–599.

Hartigan, J. (2013). Anthropology of race. In J. Hartigan (Ed.), *Anthropology of race: Genes, biology, and culture*. Santa Fe: SAR Press.

Hegeman, E., & Stocker, M. (1996). *Valuing emotions*. Cambridge: Cambridge University Press.

Olofsson, A., & Rashid, S. (2011). The white (Male) effect and risk perception: Can equality make a difference? *Risk Analysis: An International Journal*, *31*(6), 1016–1032.

Ploder, A., & Stadlbauer, J. (2016). Strong reflexivity and its critics: Responses to autoethnography in the German-speaking cultural and social sciences. *Qualitative Inquiry*, 22(9), 753–765.

Rich, A. (1984). Notes toward a politics of location. In A. Rich (Ed.), (1994). *Blood, bread, and poetry: Selected prose 1979–1985*. New York: Norton.

Said, E. W. (1978/2003). *Orientalism*. London: Penguin.
Smith, L. T. (1999). *Decolonizing methodologies: Research and Indigenous peoples*. London: Zed Books.
Spivak, G. C. (1990). *The post-colonial critic: Interviews, strategies, dialogues*. London: Routledge.
Spivak, G. (2013). *The Spivak reader: Selected works of Gayati Chakravorty Spivak*. London: Routledge.
Tsing, A. L. (2007). Becoming a tribal elder, and other fantasies of green development (reprint). In M. Dove & C. Carpenter (Eds.), *Environmental anthropology: A historical reader*. Malden, MA: Blackwell.
Tsing, A. L. (2015). *The mushroom at the end of the world: On the possibility of life in capitalist ruins*. Princeton: Princeton University Press.
Viatori, M. (2007). Zápara leaders and identity construction in Ecuador: The complexities of indigenous self-representation. *The Journal of Latin American and Caribbean Anthropology*, *12*(1), 104–133.
Viveiros de Castro, E. B. (2011). *The inconstancy of the Indian soul: The encounter of Catholics and cannibals in 16th-century Brazil*. Chicago: Prickly Paradigm Press.
Wolcott, H. F. (2005). *The art of fieldwork* (2nd ed.). Oxford: Rowman.

INDEX

Note: Page numbers followed by "n" refer to notes.